PENGUIN BOOKS

# THREE MILE ISLAND

Daniel F. Ford is an economist and writer specializing in nuclear-policy questions. A graduate of Harvard College and former executive director of the Union of Concerned Scientists, Mr. Ford has worked on a variety of studies related to nuclear-power-plant safety. He is the coauthor of six technical volumes on nuclear power and national energy policy, including *Energy Strategies* (with Steven Nadias, edited by Henry Kendall), which was named one of the best energy books of 1980 by *Library Journal*. Mr. Ford has also written articles for *The New Yorker*. He lives in Cambridge, Massachusetts.

Daniel F. Ford

# Three Mile Island

## Thirty Minutes to Meltdown

Penguin Books

Penguin Books Ltd, Harmondsworth,
Middlesex, England
Penguin Books, 625 Madison Avenue,
New York, New York 10022, U.S.A.
Penguin Books Australia Ltd, Ringwood,
Victoria, Australia
Penguin Books Canada Limited, 2801 John Street,
Markham, Ontario, Canada L3R 1B4
Penguin Books (N.Z.) Ltd, 182–190 Wairau Road,
Auckland 10, New Zealand

First published in the United States of America in simultaneous hardcover
and paperback editions by The Viking Press and Penguin Books 1982



LIBRARY OF CONGRESS CATALOGING IN PUBLICATION DATA
Ford, Daniel F.
Three Mile Island.
Includes index.
1. Three Mile Island Nuclear Power Plant, Pa.
I. Title.
TK1345.H37F67 363.1'79 81-17944
ISBN 0 14 00.6048 0 AACR2

Printed in the United States of America by
Offset Paperback Mfrs., Inc., Dallas, Pennsylvania
Set in Primer

Portions of this book appeared originally in *The New Yorker.*

*To my parents*

# Contents

# Acknowledgments

In writing this book, I have relied on data, technical analysis, editorial advice, and occasional words of general encouragement that came to me from many people. My ten years of work with the Union of Concerned Scientists provided an indispensable background on the general subject of nuclear safety. I have benefited, in particular, from a close collaboration over the years with Henry Kendall of the Massachusetts Institute of Technology, the founder of U.C.S., and to the extent that the book has a coherent outlook on the technical issues of nuclear safety, it reflects what I have learned from him. Robert Pollard of the U.C.S. staff, who formerly served on the regulatory staff of the Atomic Energy Commission and Nuclear Regulatory Commission, also provided continuing, invaluable assistance to me.

The book draws heavily on the official records compiled by the President's Commission on the Accident at Three Mile Island, the N.R.C. Special Inquiry Group, the Senate Subcommittee on Nuclear Regulation, the House Subcommittee on Energy and the Environment, and the N.R.C. Office of Inspection and Enforcement, as well as on the voluminous general documentation on nuclear-plant design and safety in the N.R.C.'s public document room in Washington, D.C. The cooperation of the officials who compiled the vast storehouse of source material on the accident is sincerely appreciated.

Portions of the book originally appeared in *The New Yorker,* and the editorial assistance of William Shawn and

C. P. Crow is gratefully acknowledged. Beatrice Rosenfeld edited the full book for Viking Penguin, and it is a pleasure to acknowledge her contribution, as well as that of Patti Hagan of *The New Yorker*'s checking department. (At one point, with the magazine's publication deadline a few hours away, the verification of one particularly obscure fact involved a race between Ms. Hagan and the computer at the Nuclear Safety Information Center in Oak Ridge, Tennessee; I am grateful for the general help of the N.S.I.C. staff, although in this instance Ms. Hagan won.) In writing the book, I was fortunate to have available a model of superb factual reportage on nuclear safety: the many articles that David Burnham provided over the years for *The New York Times.* I also had the benefit of his encouragement for my efforts, as well as the cheerful moral support of Ann Bernays, Justin Kaplan, Ann Kendall, Steven Nadis, Eileen Soskin, Eric Van Loon, and other friends.

D. F.

Cambridge, Mass.
April 1981

# Three Mile Island

# 1: Class Nine Accident

COPENHAGEN—The nuclear-power industry needs more accidents to gain the experience to prevent future mishaps, the general director of the International Atomic Energy Agency said during a visit here.

The executive, Sigvard Eklund, told a news conference: "The problem of the nuclear-power industry is that we have had too few accidents."

He added, "It's expensive, but that's how you gain experience."

—*Wall Street Journal,* February 21, 1980

A cardinal rule for the designers of commercial nuclear-power plants is that all systems essential to safety must be installed in duplicate, at least, so that if some of the apparatus fails, there will always be enough extra equipment to keep the plant under control. Federal regulations governing the industry require strict conformity to this prudent design philosophy. Even when this rule is applied, however, there is a type of accident that can jeopardize the safety of a nuclear plant. This type of accident involves what is known as a common-mode failure—a single event or condition that can cause simultaneous multiple malfunctions resulting in the major disruption of the plant's safety systems. Warnings of the potential "disastrous effect" of such common-mode failures have appeared repeatedly in the official reports of Dr. Stephen H. Hanauer, a Nuclear Regulatory Commission engineer who has been one of the federal government's chief nuclear-safety experts for the last fifteen years.

In July 1969, while Dr. Hanauer was the head of the Atomic Energy Commission's Advisory Committee on Reactor Safeguards, he sent a memo on behalf of the committee to the chairman of the commission, stating that he and his colleagues, in reviewing the design of a proposed nuclear-power plant, had observed that it would incorporate certain devices "vital to the public health and safety" that could be disabled by common-mode failures. Hanauer's memo urged that the problem be corrected while the plant was under construction, but the recommendation was never carried out. On March 28, 1979—barely three months after the plant started commercial operation—it was struck by common-mode failures that contributed to what is now regarded as the worst accident in the history of commercial nuclear power. Hanauer's memorandum, which was just one of the unheeded warnings pertaining to the plant that federal safety authorities had received over the previous decade, was entitled "Report on Three Mile Island Nuclear Station Unit 2."

Dedication ceremonies for Three Mile Island Unit 2, some ten miles southeast of Harrisburg, Pennsylvania, were held on September 19, 1978. Deputy Secretary of Energy John F. O'Leary, a leading spokesman for the Carter administration's nuclear-power policies, described the new plant on that occasion as an "aggregation of capital and patience and skill and technology" that was "sort of a miracle in many ways." According to Dr. O'Leary, who served as the director of licensing at the Atomic Energy Commission from 1972 to 1974, the plant was "a scintillating success," and he added that from this achievement "it is fair to conclude . . . that nuclear [power] is a bright and shining option for this country." Dr. O'Leary returned to Washington, D.C., after the ceremony, taking back with him a souvenir Three Mile Island Unit 2 paperweight, and for months afterward he kept on his desk a photograph of himself taken during that speech—in which, it must be noted, he did not discuss Dr. Hanauer's report of July 1969 or any of the other documents in the government's internal files that pertained to potential safety problems affecting Three Mile Island Unit 2. Nor did he make any mention of a candid and much more general private memo, dated November 1, 1976, that he himself had prepared at the request of the policy-planning staff for the incoming Carter administration. In that memo, which focused on some of the weaknesses in the federal government's handling of what he referred to as "the massive safety issues associated with the commercialization of nuclear power," O'Leary said that "the frequency of serious and potentially catastrophic nuclear incidents supports the conclusion that sooner or later a major disaster will occur at a nuclear generating facility," and that the Nuclear Regulatory Commission, "as was the case with its predecessor, the Atomic Energy Commission, has been unwilling to face up to the policy consequences of assigning high probability to a serious nuclear accident." The day after the accident, an aide noted that Dr. O'Leary's memorabilia from the Three Mile Island dedication ceremonies had been removed from his desk.

Given the regulatory lapses leading up to the accident, and the extensive warnings about the plant's safety weaknesses that had long been available to federal officials, there is some question whether the event at Three Mile Island should, in a strict sense, be called an "accident." Indeed, government records indicate that in most respects what happened was a predictable outcome of known deficiencies in the type of nuclear equipment installed there. This regulatory failure raises unsettling questions about the safety of the tens of millions of people living near the other commercial nuclear-power stations that are now operating in this country under the auspices of the Nuclear Regulatory Commission (N.R.C.), which in 1975, by act of Congress, replaced the Atomic Energy Commission (A.E.C.). The questions are not new ones: they have figured prominently in the continuing public controversy over nuclear power, and they have been the focus of research that, as a member of the Union of Concerned Scientists, I have carried out for the last ten years. The results of this research had, in fact, prompted the Union of Concerned Scientists—in a report published on January 26, 1979, two months before the accident—to recommend that sixteen plants, including Three Mile Island Unit 2, be shut down for repairs.

* * *

In the early-morning hours of Wednesday, March 28, 1979, Three Mile Island Unit 2 was functioning normally, under full automatic control. (Three Mile Island Unit 1, its sister nuclear plant, had been shut down weeks earlier for refueling and was undergoing testing before it resumed operation.) No one on duty at the plant was a qualified nuclear engineer, or even a college graduate, nor had anyone there ever received detailed technical training on how to handle complex reactor emergencies. Such training was not required by the N.R.C. and was not customarily given to reactor operators in the standard one-year training programs, conducted by the utility companies and designed primarily to instruct the operators on how to go about their routine duties. Moreover, plant supervisory personnel were also not trained on how to respond to complicated plant malfunctions. Craig Faust and Edward Frederick, both of whom had formerly operated submarine reactors for the Navy and were licensed by the N.R.C. to operate the new reactor at Three Mile Island, were on duty in the Unit 2 control room. The shift supervisor, William Zewe, another veteran of the Navy's nuclear program and a licensed senior reactor operator, was doing paperwork in his office, next to the control room. Zewe was in charge of the plant during that night's graveyard shift, 11 p.m. to 7 a.m. Two technicians on Zewe's sixteen-person overnight crew were in the basement carrying out routine maintenance on part of the main feedwater system, which provided the reactor with one of two vital supplies of cooling water. The plumbing arrangements in a nuclear plant are built to exacting standards and require exceedingly careful maintenance. Moreover, the water carried by these pipes must be continuously purified to prevent foreign matter from circulating through the system and damaging any of the equipment.

The maintenance workers, who were not required to have federal licenses or to go through any federally approved training programs, were trying to cope with a problem common to every type of plumbing system—a clogged pipe. The pipe in question was a small one coming from one of eight tanks, known as polishers, that removed impurities from the main feedwater system. The polishers themselves had to be cleaned out periodically so that the sludge accumulating inside them could be disposed of and fresh filters—tiny resin beads that had the consistency of coarse sand—could be installed. The work had been going on for about eleven hours, and the small transfer pipe, which was supposed to allow sludge from Polisher No. 7 to be flushed into a receiving tank, had become blocked. Donald Miller, one of the technicians on duty, had been trying to clear the line for about an hour by shooting in water and bursts of compressed air.

Shortly before 4 a.m., Frederick Scheimann, the Unit 2 shift foreman, went to the basement to see how work on the pipe was progressing. He discussed the problem with Miller and with Harold Farst, the other technician on duty. Scheimann climbed up on top of a larger pipe so that he could look into the polisher through a glass window. "All of a sudden, I started hearing loud, thunderous noises, like a couple of freight trains," he said later. He jumped down from the pipe, heard the words "Turbine trip, reactor trip" over a loudspeaker, and rushed to the control room. The maintenance crew working on the polisher had accidentally choked off the flow in the main feedwater system, forcing Unit 2's generating equipment—its turbine and reactor, which had been operating at ninety-seven percent of full power—to shut down. The equipment was suddenly tripped at thirty-seven seconds past 4 a.m.

Routine emergency procedures were initiated automatically. Unit 2 had been equipped with multiple automatic devices, supposedly fail-safe, that were designed to respond immediately in the event of a serious plant malfunction. These devices were linked to an electronic alert system that constantly monitored all major aspects of plant performance and triggered remedial action if it detected an abnormal condition or any signs of impending trouble—including, certainly, the failure of the plant's main feedwater system. To deal with this particular problem, the plant had been equipped with three emergency feedwater pumps that were designed to operate automatically if the main feedwater system malfunctioned. Within seconds after the equipment was shut down, the three emergency feedwater pumps went into operation, just as the reactor's designers had planned. The incident should have ended there. A few seconds into the shutdown, however, all the careful plans for assuring the safety of Unit 2 were upset.

The crisis had two immediate causes. First, the nuclear reactor—the steel boiler that holds the plant's uranium fuel—had responded violently to the shutdown of the main feedwater system. The pressure of the cooling water in the reactor had increased rapidly, since it was still being heated by the hot uranium fuel, and the pressure surge had popped open a relief valve. All this activity was predictable, but what happened then had not been predicted or allowed for: the relief valve, which should have reclosed after a few seconds, jammed open. Cooling water started to drain out of the reactor at a rate of up to a hundred and ten thousand pounds an hour, or about two hundred and twenty gallons a minute—a hemorrhaging that seriously threatened the safety of the reactor. Second, the crisis was compounded by one of those instances of carelessness that easily escape detection in a complicated machine like a nuclear-power plant. Somebody, most likely during routine tests two days before, had shut two valves in the system that piped water from the emergency feedwater pumps into the parts of the cooling system where cooling water was urgently needed. The three pumps that were expected to cool the reactor—a single steam-driven emergency feedwater pump and two emergency feedwater pumps powered by electric motors—were rendered completely ineffective. In other words, a testing procedure had created a common-mode failure—a multiple safety-system failure—of the sort that, because of its supposedly remote probability, had been officially regarded as an "incredible" event.

During the first few minutes of the accident, conditions at the plant were grave but not desperate, since there was a remaining set of safety devices available to help cool the reactor. This equipment consisted of three additional pumps—part of the high-pressure injection system, which was connected to a special reservoir of emergency cooling water—that were capable of resupplying the reactor with a thousand gallons of water a minute, or more than enough to compensate for what was being lost through the stuck-open relief valve. Two minutes after the accident began, these emergency pumps were automatically turned on, but to no avail, for the reactor operators monitoring the accident on instruments in the control room decided to shut them off. They throttled the pumps so that practically no emergency water could be delivered to the reactor. This shutoff occurred some four minutes after the accident began, and represented a further common-mode failure—in this case, a deliberate maneuver by the operators to override the plant's automated safety equipment. The operators then inadvertently worsened the accident still further. They opened a drain line to remove even more water from the reactor—an action that effectively doubled the severity of the coolant loss. Faust, Frederick, Zewe, and Scheimann, who were trying to cope with the accident as best they could and were performing in accordance with their limited training, had interpreted control-room instrument readings to mean that the reactor had too much water in it rather than too little. They were unaware of the open relief valve.

The control room at Unit 2 had no indicator that directly reported whether or not the relief valve was open, and it remained open, unknown to the operators, for some two hours and twenty minutes. (Brian Mehler, the supervisor of the incoming shift, arrived early that morning and surmised that the relief valve was open from his review of plant conditions.) Nor—odd as it may seem—was the control room equipped with any instrument that directly told the operators how much cooling water was inside the reactor. Instead, the operators had been trained to monitor the water level in another tank—a separate one, called the pressurizer—which was connected to the reactor's primary cooling system. The pressurizer, which was partly filled with hot water, helped control the pressure of the reactor's primary cooling water. The operators had been instructed, in the absence of a reactor-water-level indicator, to infer the amount of water in the reactor from the amount in this auxiliary tank. Accordingly, since the pressurizer instruments indicated that the pressurizer was full, the operators assumed that the reactor was full, too. This particular way of interpreting their instruments, Frederick later explained, was "what everybody punches into you" during the operator training program.

Within five minutes after Unit 2's main feedwater system failed, the reactor, deprived of all normal and emergency sources of cooling water, and no longer able to use its enormous energy to generate electricity, gradually started to tear itself apart. The pressure of the water inside, which had increased suddenly in the few seconds after the accident began, now kept decreasing uncontrollably. The water remaining inside the reactor began to flash into steam, which in the next few hours expanded and blanketed much of the reactor's fuel rods, preventing effective cooling. A rash of further instrument problems, equipment malfunctions, computer-system breakdowns, and other difficulties beset the Unit 2 operators and a growing number of plant officials who were coming to their aid. "It seemed to go on and on, surprise after surprise," Thomas Mulleavy, a radiation-protection supervisor at the plant, has recalled. "The equipment that we had to use did indeed malfunction, as most equipment will do on occasion, and always seems to when you need it most." At 7:24 a.m., shortly after Gary Miller, the station manager, arrived in the Unit 2 control room, a "General Emergency" was declared—the first ever to arise at a commercial nuclear-power plant in the United States.

The net result of a long chain of human and mechanical failures was that for some sixteen hours the hot uranium-fueled core in the Unit 2 reactor was not adequately cooled. The uranium fuel rods overheated, swelled, and ruptured, according to postaccident N.R.C. estimates, with about a third of the core reduced to rubble. Large amounts of radioactive material were released from the damaged fuel rods, and, because of the open relief valve, much of this material escaped into the containment building housing the reactor. The atmosphere there became what one N.R.C. official later described as "murderously radioactive," and thousands of gallons of radioactive water from the reactor were accidentally pumped from the containment building into a less secure auxiliary building. Plant officials, however, assured the public throughout the day that there was no threat of a radiation release. From the data the agency received from the plant, the

N.R.C. concluded that the reactor had been kept in a stable condition, and in very carefully worded public statements agency officials went to great lengths to play down the seriousness of the accident. The N.R.C. chairman, Joseph Hendrie, briefing members of Congress the next day—March 29—said that there might have been some minor "cracks" in "perhaps about one percent" of the reactor's uranium fuel rods, but he emphatically assured them that there was no serious "ongoing problem" at the plant. Instrument readings from the plant by then showed extremely high radiation levels in the containment building, but Hendrie dismissed these readings as an "oddball" instrument error.

Not until early on the morning of Friday, March 30—two days after the start of the accident—did the magnitude of the problem begin to be understood by plant and N.R.C. officials. A three-and-a-third-ounce sample of cooling water taken from the reactor late Thursday afternoon was found to contain so much radioactive material that it became clear an appreciable fraction of the reactor's fuel rods had been grossly damaged. N.R.C. analysts admitted to being dumbfounded by "failure modes" that "had never been studied," and they concluded that the fuel damage might make it extremely difficult to keep the reactor safely under control. As the officials pondered the difficulties, they received word from the plant that unexpected bursts of radioactive gases had been released from Unit 2 and were being blown by the wind toward some of the neighboring towns. Furthermore, N.R.C. officials, relying on hasty consultations with expert advisers, concluded that a large, growing, and potentially explosive "hydrogen bubble" might have formed inside the reactor—a development that conjured up the specter of a possible large-scale release of radioactive material from the plant. Federal and state officials began preparing contingency plans for an evacuation of the area, and announced to the public that evacuation might be necessary. An estimated hundred and forty thousand people fled the area during the weekend—some in response to advice issued by Pennsylvania's Governor Richard Thornburgh to pregnant women and to families with preschool children, but most through their own good sense.

By the evening of Sunday, April 1, the atmosphere of crisis had begun to abate, for the N.R.C. had found that its fears of a hydrogen explosion were based on spurious technical analysis. There was never any possibility of a hydrogen explosion, the agency learned, to its embarrassment, because the amount of oxygen needed to combine with the hydrogen to make an explosive mixture was simply not present in the reactor. Moreover, the hydrogen in the reactor could be removed by relatively straightforward techniques. This encouraging news still left N.R.C. and Unit 2 officials facing several pressing problems, however, the most important of which was trying to keep the uranium fuel in the reactor adequately cooled. To aid in this effort, industry experts from around the country converged on Three Mile Island and formed themselves into expert advisory groups, which developed plans for providing long-term cooling of the reactor and for preventing further leakage of radioactive materials from the containment and auxiliary buildings. On April 27, after weeks of delicate maneuvering, the plant was finally stabilized and put into controlled shutdown conditions.

Some three years after the accident, Unit 2 remains shut down. Its uranium fuel has now been much cooled off, but its containment building and auxiliary tanks are still flooded with almost a million gallons of radioactive water. The only work in progress on the island is a cleanup effort aimed at disposing of the radioactive debris left from the accident. Metropolitan Edison, the company that operated the plant, has recruited a work force for this grim chore, which is expected to go on for years and to cost at least a billion dollars. The company's employment advertisements, such as one placed in *The New York Times* in late 1979, promised prospective recruits for the cleanup work "unprecedented opportunities for new scientific experience" and "immediate ground floor opportunities for dedicated scientists and engineering personnel who want to be in the forefront of emerging technologies."

* * *

Unit 2, before its operation was interrupted on March 28, 1979, was performing a routine task that has been of growing importance to society since the early eighteenth century: it was boiling water to produce steam. Since 1711, when an ironmonger named Thomas Newcomen devised a steam engine to pump water out of the English coal mines, steam power has been a decisive factor in the world's economic progress. Several advances introduced later in the eighteenth century by James Watt, an instrument maker from Glasgow, led to his invention of a rotative engine powered by the expansion of steam. Watt's steam engine became the source of power for new techniques of spinning and weaving cotton in the late eighteenth century. These separate technical innovations combined to produce the textile factories that transformed agrarian England into an industrialized nation. Steam subsequently became the motive power for railroads and shipping, and by the end of the nineteenth century it had begun to spin turbines, invented for the generation of electricity.

In this century, steam boilers of increasing size and sophistication—fired, successively, by wood, coal, oil, natural gas, and nuclear energy—have been developed. Today, a single large generating station can supply all the electricity needed by hundreds of thousands of consumers. A concerted effort over the last century by private corporations and, more recently, the federal government has carried forward the development of technology for central generating stations. Among the important innovators in boiler technology was a Rhode Island engineer named Stephen Wilcox, who in 1856 proposed a new type of water-tube boiler, capable of producing steam more efficiently than existing boilers and with less risk of a boiler explosion. Joining forces a decade later with a New York inventor named George Herman Babcock, Wilcox founded a company, which patented its first boiler in 1867 and continued to develop boilers of advanced design. The Babcock & Wilcox Company, which is now a subsidiary of McDermott, Inc., and has become a major multinational enterprise, is one of the four principal American manufacturers of nuclear reactors. (The others are the Westinghouse Electric Corporation, Combustion Engineering, and General Electric.) Two of the nine Babcock & Wilcox reactors currently licensed for commercial operation in the United States are on Three Mile Island, and Babcock & Wilcox reactors are on order for ten more nuclear plants, in the United States and abroad.

Though comparable in its basic purpose to the brewer's copper kettle that was used as a boiler in Thomas Newcomen's revolutionary steam engine, a nuclear steam-supply system is a machine of the utmost complexity. The centerpiece of the system is a reactor vessel that is connected to one of the most elaborate plumbing systems ever devised. The typical Babcock & Wilcox reactor, to use a specific example, is a steel bottle that stands some forty feet high, measures fifteen feet across, and weighs more than four hundred tons. Inside the reactor's steel walls is the uranium core, itself weighing a hundred tons or more, where controlled chain reactions provide the heat that, as in a conventional electric plant, is used to turn water into steam to drive a turbine. The reactor's core is impressively compact: it is only twelve feet wide and twelve feet high.

In a controlled nuclear reaction, uranium nuclei fission—split apart—with a consequent release of thermal energy, which can be used to convert water into steam to power a turbine. It was not until the end of 1938, when Otto Hahn, Fritz Strassmann, Lise Meitner, and Otto Frisch discovered how the uranium nucleus could be fissioned, that any method for tapping the energy inside the atomic nucleus was even theoretically available. The successful theoretical and experimental work carried out during the Second World War under the Manhattan Project led to the development of methods for fissioning a relatively rare form of uranium—U-235. The knowledge gained about how to split the nucleus of U-235 was used—as all the world is aware—to develop weapons of unprecedented destructive ability. After the war, atomic-bomb scientists advocated the commercial production of electric power by a domesticated version of the process that had devastated Hiroshima and Nagasaki. For some thirty-five years, the United States government, several major private companies, and cadres of technical specialists have worked to exploit nuclear energy as a means of powering electric-generating stations. Over most of this period, the promoters of the program, the American scientific establishment, and many members of the general public have looked upon uranium as the basic fuel that the nation will depend on for its future material prosperity.

In most modern power reactors, the uranium fuel is in the form of millions of ceramic uranium-dioxide pellets, each about the size of the tip of one's little finger, which are stacked inside slim tubes made of an alloy called Zircaloy. These fuel rods—there are about forty thousand of them in a typical Babcock & Wilcox reactor—stand upright inside the reactor, as if someone had created a dense forest of precisely spaced metal trees. Putting the uranium fuel in large numbers of metal tubes rather than in one clump increases the surface area, so that the water flowing through the reactor can more effectively carry off the heat generated by fission. This is important for both safety and economic reasons: the arrangement helps prevent the reactor fuel from overheating, and allows the most efficient use of a given quantity of fuel.

There are several types of commercial power reactors. Some use ordinary water to cool their fuel, some are cooled by a gas, such as carbon dioxide, and some use liquid metal as a coolant. A Babcock & Wilcox reactor is generically identified as a pressurized-water reactor—a type of reactor that was developed to power nuclear submarines for the Navy and has become the world's most widely used commercial power reactor. There are now forty-three pressurized-water reactors licensed to operate in the United States, and sixty-four more are operating around the world. In a pressurized-water reactor, the water is kept under extremely high pressure—approximately twenty-two hundred pounds per square inch—so that it can absorb the heat from the nuclear reaction and yet not turn to steam. The pressurized water, known as the primary cooling water, must be kept moving through the reactor—or else it would soon overheat—and large, nine-thousand-horsepower pumps are provided to keep it flowing upward through the hot uranium-fueled core. After it emerges from the core, the heated water flows through pipes three feet in diameter into steel tanks—the steam generators—that are partly filled with water supplied by the plant's main feedwater pumps. The water supplied by the main feedwater pumps is designated the secondary cooling water. The heated water from the reactor does not mix with the water supplied to the steam generators by the feedwater pumps; instead, the water from the reactor flows through thousands of small tubes inside each steam generator. These tubes are surrounded by the water—the secondary cooling water—that is supplied by the feedwater pumps. Much as a conventional radiator for home heating passes hot water through its coils to warm the surrounding air, heated water from the reactor, moving through the tiny steam-generator tubes, heats the colder secondary cooling water surrounding the tubes. Like a relay race, in which one runner passes a baton to another, the reactor-cooling system is designed so that the heat added to the primary cooling water as it passes through the core is subsequently transferred—inside the steam generators—to the secondary cooling water.

The water in the secondary cooling system, which is kept at much lower pressure than the primary cooling water, boils and turns into steam, and this steam, as in a conventional power plant, is piped into a turbine, where it expands and, in so doing, spins the large turbine blades, the way wind turns a windmill. The shaft of the turbine is connected to a giant electric generator, and the turning of the shaft spins the generator. The net result of all this motion is an electric current that can be sent through transmission lines to the power station's customers.

If all goes well, the reactor and the steam generators in a nuclear-power plant of the pressurized-water variety maintain a stable, businesslike relationship such as might obtain between two complementary monopolies. The reactor can be thought of as selling heat to the steam generators. The various water-circulating systems in the plant provide the transportation for this commodity: the primary cooling water delivers heat from the reactor to the steam generators, and the secondary-cooling-water system delivers the heat—in the form of high-pressure steam—from the steam generators to the turbine. In terms of the conventional power-plant goal of using steam to rotate a turbine generator, the power-producing achievement of the primary and secondary cooling systems in a typical Babcock & Wilcox nuclear plant would surely astonish James Watt. Every hour, the steady working relationship between the reactor and the steam generators in one of these plants creates more than eleven million pounds of high-pressure steam. This steam, which causes a giant turbine generator to make eighteen hundred revolutions a minute, produces about nine hundred thousand kilowatt-hours of electric power.

For a nuclear plant incorporating a pressurized-water reactor to function satisfactorily, its internal energy supply and demand must neatly balance: the output of heat by the reactor has to match the uptake of heat by the steam generators. If this delicate equilibrium is upset, the temperatures and pressures inside the reactor can increase or decrease suddenly and create an accident in which the uranium fuel can overheat, perhaps uncontrollably. One obvious direct threat to normal cooling is the rupture of any of the pipes that deliver primary cooling water to the reactor. A more indirect but no less threatening interference with the reactor-cooling process can arise from malfunctions in the secondary cooling system, such as a failure in the main feedwater pumps that supply the steam generators. In that circumstance, with the steam generators no longer functional, there would be no method for removing the heat generated by the uranium fuel, and the fuel would overheat. Furthermore, the relationship between the primary and secondary cooling systems is so complex that there are circumstances in which too much cooling water from the main feedwater pumps would be as much of a threat as too little. An excessive supply of secondary cooling water to the steam generators could drain so much heat out of the reactor that the pressure of the primary cooling water might suddenly decrease. If that occurred, some of the primary cooling water could flash into steam, and the resulting steam pockets in the reactor could prevent needed cooling water from reaching the uranium fuel and keeping it adequately cooled. Increases and decreases—large ones, especially—in the temperature and the pressure of reactor-cooling water must be carefully avoided in order to maintain steady power production as well as safe and stable cooling of the reactor's fuel. This requirement must be given priority by both plant designers and plant operators.

* * *

To keep a nuclear reactor operating smoothly and under control, a number of systems, subsystems, components, structures, and people have to work together in a coordinated and reliable way. The task is customarily accomplished by an operating crew of a few dozen people with the aid of—at current prices—up to a billion dollars' worth of hardware. The focus of the plant operators, and the purpose of much of the elaborate machinery they work with, is the absolutely essential task of keeping the power levels, pressures, and temperatures inside the nuclear reactor completely under control. To this end, the plant is equipped with intricate systems linked by miles of electrical cables—the central nervous system of the plant. Data from electronic monitors are fed into ranks of computers and automated protection systems that, in theory, will respond swiftly and appropriately to any contingency that might compromise the reactor's steady operation. Federal safety regulations, rules, and standards set forth—sometimes precisely, but sometimes only in a general fashion—the precautions required of each commercial nuclear generating plant. All components, systems, and structures important to safety and all operator actions and procedures that could affect the health and safety of the public are supposed to be covered by the web of federal nuclear-safety regulations. The plants must be equipped with vast arrays of safety equipment. This equipment must be highly reliable—conservatively designed, routinely tested, carefully maintained, and usually installed at least in duplicate, so that redundant equipment will compensate for the failure of a single system or component. Controlling the power level of the reactor, for example, is a critically important safety priority, and highly reliable methods for rapidly inserting control rods to shut off the nuclear reaction are mandated. Sixty-nine control rods are installed in the typical Babcock & Wilcox reactor for this purpose. The control rods act as blotters to absorb neutrons—the subatomic particles that are given off when a U-235 nucleus fissions and that then hit the nuclei of other U-235 atoms and cause subsequent fissions. By capturing neutrons, and so curtailing the fission chain reactions, these rods control the reactor's power output. Cooling the reactor is another major focus of the regulatory framework governing the nuclear industry. Plants are required to have several networks of devices that can propel supplies of cooling water through the reactor during all the different types of cooling difficulties that can arise.

To generate power and comply with the safety guidelines, a nuclear-power station typically requires some fifty miles of piping, held together by twenty-five thousand welds; nine hundred miles of electrical cables; eleven thousand five hundred tons of structural steel; and a hundred and seventy thousand cubic yards of concrete. Countless electric motors, conduits, batteries, relays, switches, switchboards, condensers, transformers, and fuses are needed. Plumbing requirements in the various cooling systems call for innumerable valves, seals, drains, vents, gauges, fittings, pipe hangers, hydraulic snubbers, nuts, and bolts. Structural supports, radiation shields, ductwork, fire walls, equipment hatches, cable penetrations, emergency diesel generators, and bulkheads must be installed. Instruments must be provided to monitor temperatures, pressures, chain-reaction power levels, radiation levels, flow rates, cooling-water chemistry, equipment vibration, and the performance of all key plant components. Written procedures must be provided to cover normal operations, equipment installations, periodic maintenance, component and system testing, plant security, and appropriate operator actions during reactor startup, reactor shutdown, and all anticipated emergencies.

All nuclear-power-plant systems, structures, components, procedures, and personnel are potential sources of failures and malfunctions. Problems can arise from defects in design, manufacturing, installation, and construction; from testing, operational, and maintenance errors; from explosions and fires; from excessive corrosion, vibration, stress, heating, cooling, radiation damage, and other physical phenomena; from deterioration due to component aging; and from externally initiated events such as floods, earthquakes, tornadoes, and sabotage. The possibilities of such failures and their consequences are supposed to be studied carefully before nuclear plants are licensed to operate, and appropriate steps are supposed to be taken to prevent any failures that would lead to major nuclear-radiation accidents. Over the years, government safety analysts have looked into a large number of possible accidents that could arise at nuclear plants, and they have made judgments about which types of possible accidents have a high enough probability of occurring to pose a "credible" threat to plant safety. To streamline the safety-review process, accidents presumed to be "credible" have been divided into eight categories, ranging from trivial mishaps, known as Class One accidents, all the way up to major disruptions of the plant—Class Eight accidents, which would require the emergency operation of plant safety systems. A.E.C. and N.R.C. safety reviewers have concluded that even in the worst Class Eight accidents, plant safety systems would satisfactorily prevent any serious damage to the reactor's core and any serious release of radioactive materials into the environment. It is possible, they admit, that there are "hypothetical" circumstances in which the safety apparatus could fail, thereby setting the stage for a huge release of radioactive debris into the neighboring region. Still, these potentially catastrophic accidents have been assumed to be so unlikely that they pose no "credible" safety problem. All such possible but "incredible" accidents—accidents that supposedly could never happen—are lumped together, for bureaucratic convenience, in one catchall category. In the official shorthand, they are simply referred to as Class Nine accidents.

Experience with accidents in complex systems shows that major mishaps often have humble beginnings, easily overlooked even by diligent safety reviewers. Simple malfunctions in a system as subtle and intricate as that of a nuclear reactor can combine with unsuspected flaws or induce other malfunctions in some vulnerable piece of equipment. As the chain of malfunctions proceeds, circumstances may require the emergency operation of one or more of the nuclear plant's safety systems. Such a scenario presupposes, of course, that the contingency is one for which plant designers have provided appropriate safety apparatus. If not—if the accident unfolds in a way against which no protection has been provided—the accident may cause the plant to go out of control. Safety systems, even if they are incorporated into the plant design, may or may not perform as intended. The accident may be terminated safely or it may develop in ways that can set the stage for a terrible nuclear-radiation calamity.

The most widely feared type of reactor accident is one that involves the overheating and melting of the reactor's uranium fuel. As a reactor operates, the fissioning of uranium nuclei results in the accumulation of radioactive wastes—nuclear ashes, so to speak—inside the uranium fuel rods. These wastes, principally the fragments of the split uranium nuclei, give off an amount of radiation that itself constitutes a significant source of heat. In normal operation, this so-called decay heating is equivalent to six or seven percent of the full power of the reactor. The radioactive processes responsible for the persistent decay heating cannot be controlled, as the chain reaction can. Thus, even when the fissioning of the uranium fuel is halted—which occurs when the control rods are inserted in the reactor's core—the reactor continues to generate heat. The heat produced by the radioactive decay does diminish with time—rapidly at first, then more slowly—but for months after a reactor is shut down it remains at levels that require continuous cooling-water circulation.

Inadequate core cooling can give rise to what is termed a meltdown accident. If, for example, the reactor suddenly lost its primary cooling water through a large pipe rupture and was not quickly supplied with emergency cooling water, the temperature of the fuel rods would rise from the normal six hundred degrees to more than thirty-three hundred degrees, at which point the Zircaloy tubes that hold the uranium would begin to melt, causing the reactor core to lose its precisely arranged geometry. The uranium itself, heated by the radioactive waste materials that had been generated during reactor operation, would in turn begin to melt when the temperature reached about five thousand degrees. In short order, the core would turn into a white-hot blob of molten radioactive metal. Pouring to the bottom of the reactor, the fuel could melt its way through the steel reactor vessel in half an hour, and would then drop onto the thick concrete floor of the containment building. The containment building is not designed to withstand a meltdown; it has no special features to stop the molten fuel from penetrating the concrete and continuing downward into the ground. (The assumption in designing containment buildings has been that meltdown accidents have a negligible probability.) The sequence of events in a meltdown accident has been called, as long as anyone in the business can remember, the China Syndrome—a facetious reference to the general direction in which the molten core would be traveling. If a meltdown accident took place, the containment building could undergo several forms of collateral damage. The buildup of pressure, the formation of hydrogen and carbon-dioxide gases, explosive reactions between molten metal and residual water in the building, and other phenomena could damage the containment building and allow radioactive contaminants to escape into the surrounding area.

The radioactive material in a typical commercial reactor represents an almost unimaginably large quantity of biologically hazardous material. Using the unit of measurement named after Marie Curie, a leading pioneer in nuclear physics, scientists describe the typical nuclear plant's normal accumulation of radioactive materials as about fifteen billion curies. In terms that may better illustrate the potential hazard, it can be said that the long-lived radioactive material in a modern commercial reactor would approximate the long-lived radioactive fallout produced by the detonation of more than a thousand nuclear weapons equivalent in size to the bomb dropped on Hiroshima. Though there is no possibility of a nuclear explosion at commercial reactors, a simple leak of any appreciable portion of a plant's radioactive materials is all that is necessary to set the stage for a major catastrophe. Some of the accumulated radioactive material is gaseous or volatile in form; this material, which is normally locked inside the uranium-fuel pellets, would be released in a meltdown. Various paths out of a damaged nuclear plant could allow an invisible, lethal cloud of radioactive gases to be blown across the neighboring countryside.

Several estimates have been made concerning the consequences of meltdown accidents. These consequences depend on population density near the plant, prevailing weather conditions, and the effectiveness of any evacuation that can be arranged, as well as on the circumstances of the accident itself. In "fortunate" cases—a seaside reactor and just the right wind direction—the neighboring population might be spared immediate injury, though the molten fuel, embedded in the ground under the plant, could leach large quantities of radioactive contaminants into the groundwater and neighboring bodies of water for decades. In less fortunate cases—for instance, a large release of radioactive gases upwind of a populated area under adverse weather conditions, such as a temperature inversion, which would keep the radioactive cloud trapped close to the ground—widespread injuries could result. These would include both short-term, life-threatening injuries from acute radiation exposure and long-term increases in the incidence of cancer and genetic defects in the exposed population. Ingestion and inhalation of radioactive materials would also contribute to the exposed population's radiation dose. According to the estimates published in an official government study of reactor safety completed in 1975, a major nuclear accident could result in thirty-three hundred immediate fatalities, forty-five thousand cases of acute radiation injury, and fourteen billion dollars in property damage. Longer-term health effects from the same hypothetical accident were estimated to include forty-five thousand latent cancer fatalities and forty thousand cases of thyroid tumors (caused by exposure to radioactive iodine, which would be released from the damaged reactor and concentrated in the thyroid gland after it was inhaled or ingested), and five thousand genetic defects in the first generation after the accident. Another government study, which was completed in 1965 by the A.E.C. but never published, concluded that a major reactor meltdown accident could create an "area of . . . disaster" that "might be equal to that of the State of Pennsylvania." The hypothetical reactor that could produce such an accident was a pressurized-water reactor only slightly larger than the one at Three Mile Island Unit 2. On March 28, 1979, during the critical early phase of the Three Mile Island accident, Unit 2—according to engineering estimates prepared by a special N.R.C. study group—came within thirty to sixty minutes of a meltdown.

The near-meltdown of Unit 2 represented a Class Nine accident—a supposedly "incredible" event that prompted Dr. Hanauer, a few days later, to issue a memo to top N.R.C. officials in which he announced certain "changes in my thinking." He wrote, "Core damage is credible." This new outlook on Class Nine accidents was also noted a few months later in a speech by Harold Collins, the N.R.C.'s assistant director for emergency preparedness. He remarked that as a result of the accident the "cherished" official notion that "the chances of a serious accident occurring were extremely remote . . . has, in my view and the view of others, been essentially 'knocked into a cocked hat.' "

* * *

Three Mile Island is a flat area of some four hundred acres, with rich, sandy silt underlaid by Gettysburg shale. It lies about nine hundred feet from the east bank of the Susquehanna River and ten miles southeast of Harrisburg, Pennsylvania's capital. The west bank of the river is just over a mile away, and about a mile and a half south of the island is the York Haven Dam. Three Mile Island is in Dauphin County, in the middle of the heavily industrialized region of south-central Pennsylvania. The county has a population density seven times the national average, with the highest concentrations northwest of the island along the east bank of the river, including Harrisburg and such smaller municipalities as Steelton, Highspire, Middletown, and Royalton. There has always been vague official recognition of the fact that prudence requires the siting of nuclear plants at reasonable distances from populated areas, to minimize the number of people who might be injured by potential accidents, yet numerical criteria specifying the maximum allowable population density in the vicinity of a proposed plant site have never been issued by federal nuclear-safety authorities. In the late sixties, when the Atomic Energy Commission approved the construction of the Three Mile Island plant, there were six hundred and twenty-one thousand people living within twenty miles of the site.

Until recent years, Three Mile Island was actively farmed, like much of the land in Dauphin County and in neighboring York and Lancaster counties. Some land in those counties is still used for dairy farming and poultry farming and for growing tobacco, vegetables, fruit, alfalfa, corn, and wheat. A farmer who leased two hundred and seventy acres of Three Mile Island in the 1950s and 1960s used the land primarily to grow corn and tomatoes, and since there was no bridge to the island, he transported his equipment and produce by barge. Seventy vacation cabins were built on the island, nestled in among its woodlands. As in many a near-paradise, there was luxuriant poison ivy and no electricity.

Since early in this century, Three Mile Island has been owned by public-utility companies—the General Public Utilities Corporation and its predecessors—which leased portions of it for farming and recreational use. G.P.U. is a holding company based in Parsippany, New Jersey. Three G.P.U. operating subsidiaries—Metropolitan Edison, the Pennsylvania Electric Company, and the Jersey Central Power & Light Company—serve a total of more than one and a half million customers in Pennsylvania and New Jersey. In 1966, G.P.U. announced that it wanted to use Three Mile Island to further a regional power-development plan by making it the site of a commercial nuclear-power station. The application to construct Three Mile Island Unit 1 was filed with the A.E.C. on May 1, 1967. The proposed plant was to be operated by Metropolitan Edison. The company also decided that it would add a second unit to its Oyster Creek Nuclear Station, in Oyster Creek, New Jersey, which would be operated by the Jersey Central Power & Light Company. A construction permit application for this plant was filed on April 29, 1968. In January 1969, however, because of labor problems affecting the New Jersey site, G.P.U. announced that the additional plant intended for Oyster Creek would instead be built as Three Mile Island Unit 2. It would be owned jointly by the three G.P.U. subsidiaries and operated by Metropolitan Edison. The A.E.C. processed the applications for the two Three Mile Island nuclear plants expeditiously, held brief public hearings on the applications, and issued the appropriate construction permits. The permit for Unit 1 was issued on May 18, 1968, and the one for Unit 2 on November 4, 1969. The A.E.C.'s regulatory staff wrote an official safety-evaluation report for each plant stating that there was "reasonable assurance" that the proposed plant could operate "without undue risk to the health and safety of the public."

There was nothing about the design of either Unit 1 or Unit 2 that specially prompted a broad and reassuring official conclusion about the plant's prospective safety. The A.E.C., relying on the good sense and good faith of what it regarded as "self-regulating" companies rather than on thorough independent checking and detailed federal safety instructions, had issued dozens of nuclear-plant-construction permits before those for Three Mile Island. When these two plants were proposed, the A.E.C., without pause for reflection, followed its regular procedure and awarded its official safety imprimatur almost automatically. After all, the A.E.C. had set up such cooperative licensing arrangements with the emerging nuclear industry because, under the terms of the Atomic Energy Act of 1954, the agency's function was to promote as well as to regulate nuclear energy. This dual and conflicting mandate, if adhered to scrupulously, might have created a continuing struggle between the A.E.C.'s ambitions and its prudence. Things hardly came to that, however, because from its earliest days the promotional role dominated all other A.E.C. interests and obligations. It was easy to approve the Three Mile Island Nuclear Station, since this facility could be treated as just one of a batch of Babcock & Wilcox nuclear plants, whose construction had already begun with A.E.C. approval.

When Unit 1 was completed in 1974, the A.E.C. issued a federal operating license for the facility. Then, in 1978, when work on Unit 2 was finished, the N.R.C., having replaced the A.E.C. as the federal nuclear licensing authority, issued the formal certificates permitting Unit 2 to go into commercial service. N.R.C. officials were aware of still unresolved issues—noted in private A.E.C. files, such as Dr. Hanauer's memorandum of July 1969, on the problem of potential common-mode failures—that materially affected the safety of this plant and others authorized for construction by its predecessor agency. Before licensing Unit 2, N.R.C. experts had flagged some fourteen "open safety items"—problems needing further technical evaluation—including one that dealt with the ability of the plant's safety apparatus to control small loss-of-coolant accidents (such as can be created by a stuck-open relief valve). But the N.R.C., carried forward by the A.E.C.'s momentum and by the commitment of successive administrations to nuclear expansion, licensed Three Mile Island Unit 2 anyway. According to Arizona Governor Bruce Babbitt, who heads a new presidential Nuclear Safety Oversight Committee that reviews N.R.C.'s performance, Unit 2 was licensed on the basis of the "unquestioned assumption by the N.R.C. . . . that any utility that wanted to produce nuclear power could do so—a policy that no matter how small or unsophisticated the utility, it was eventually entitled to wrap its arms around a nuclear reactor."

There had been little local opposition to the proposal to convert Three Mile Island into a nuclear generating complex. The Metropolitan Edison Company had adopted a course that has become customary for American utility companies entering the nuclear-power business: it had pursued a low-key but aggressive public-relations strategy. Local officials were taken on tours of the site and provided with data emphasizing the economic advantages that would accrue to the region as a result of the project. Local newspapers were provided with the same information, and they approvingly passed it on to their readers. One strong selling point was the boost that the plant-construction activities would give local business, the housing market, and even the school system; the construction, it was said, would help the area recover from the closing, in 1964, of the nearby Olmsted Air Force Base, which had been one of the largest employers in the region. And, indeed, in 1972, at the peak of construction on Three Mile Island, the project employed a work force of thirty-one hundred and twenty. The local Chamber of Commerce was persuaded without difficulty of the merits of the nuclear plant, and local unions provided further support for the project, which was entirely a union job. Met Ed obligingly relocated the existing Three Mile Island summer cottages on other islands in the river, and proposed an extensive recreation complex on Three Mile Island itself as a further spur to local acceptance. People in the communities abutting Three Mile Island expressed little concern about their safety, and over the years they received mostly comforting news on the subject from the local newspapers, whose reporters had benefited from information supplied by Metropolitan Edison. (A feature article that appeared in the Harrisburg *Evening News* on January 14, 1969, for example, had the headline "ON THREE MILE ISLAND: NUCLEAR REACTOR IS NOTHING TO GET 'STEAMED UP' OVER." Such articles were highly effective in reassuring the local population; this one, by Alec Green, pointed out that one of the main materials used in the plant was boric acid, which, it said, was "just plain eyewash.") The few citizens who worried about the plant—about, for example, possible aircraft accidents caused by the proximity of the plant's large cooling towers to Harrisburg International Airport (formerly Olmsted Air Force Base), about the effects of low-level radiation from the plant, and about the lack of adequate evacuation plans in the event of an accident—were unable to muster the financial resources or the political clout needed to mount an effective challenge to the company. Efforts by these citizens to press their concerns through legal action did not succeed in preventing the construction and operation of the facilities, nor was community sentiment in favor of the plant greatly altered by the efforts of the local activists.

The only major difficulties slowing construction of the plant arose from brief strikes; from Hurricane Agnes, and consequent flooding, in 1972; and from financing problems that the company experienced as it tried to meet the increasing costs of the project. Unit 1, which had been expected to cost a hundred and ten million dollars, was completed in 1974 at a cost of nearly four hundred million. Unit 2, whose completion was delayed several times by lack of funds, was supposed to cost a hundred and thirty million dollars, but by the time it went into commercial operation, on December 30, 1978, its total cost was estimated to be more than seven hundred million. The combined bill for the project was thus one billion one hundred million dollars. Cost escalation of this magnitude has been a recurring problem for the commercial nuclear power industry in the United States.

* * *

The high cost of building large nuclear power stations such as Three Mile Island Unit 2 has had a pervasive impact on the safety of these facilities, and it played an important role in bringing about the events of March 28, 1979. With their high fixed costs, the only way nuclear plants can produce power economically is by operating as close to a hundred percent of their capacity as possible for as much of the year as possible. On-again, off-again operation that leaves a nuclear plant idle much of the time vastly increases the cost of the electricity it generates because the fixed costs have to be spread over a smaller output of electricity. Accordingly, utility companies want their nuclear plants operating at high capacity to provide continuous "baseload" electric production for their customers; the companies rely on less expensive conventional power plants to meet daily and seasonal periods of peak demand.

In the effort to keep their nuclear units at full power day in and day out, except for an annual refueling shutdown of a few weeks, utility companies frequently find themselves under intense pressure to compromise on safety. To acknowledge safety problems would be to expose themselves to mandatory equipment changes or repairs that could mean extensive plant shutdowns; their tendency, therefore, is to discount evidence of possible safety problems and to keep their nuclear units on line. Even when equipment defects or safety difficulties are acknowledged, there is a strong financial incentive for the companies to try to delay necessary repairs that would require the plant to cease operation, even temporarily. The established practice is to seek to put off major repairs until the next scheduled refueling shutdown, which might be several months away—or, if federal officials permit, to postpone the repairs still longer. The solutions to newly discovered safety problems are not always immediately obvious, and the companies are naturally unwilling to keep their expensive nuclear plants out of service for an indefinite period—it might stretch into years—while research and testing are undertaken to find answers to intricate safety questions. Thus, although the state of the art of nuclear safety has changed very rapidly in the last decade, there has been an underlying financial obstacle to the introduction of improved safety technology: not only the direct costs of new devices but also the costs that would arise from what might be a prolonged shutdown while the new equipment was installed or some part of the plant rebuilt. Utility companies have thus steadfastly opposed any general policy that would require them to upgrade their nuclear plants on a regular basis to keep them supplied with the best available safety technology. The A.E.C., in the interests of protecting the economic viability of nuclear electric-power production, routinely exempted nuclear plants already built from meeting subsequently developed federal safety standards, and the N.R.C. retained this forgiving attitude. As a result, new federal regulations almost always contain a grandfather clause that allows existing plants to continue in operation without the safety design changes or improvements mandated for new plants. This policy, in the view of some of the government's senior safety analysts, has led to the accumulation of increasing numbers of operating nuclear plants with pronounced safety defects.

The safety compromises made by the Metropolitan Edison Company and permitted by federal safety policies formulated to encourage the expansion of the commercial nuclear industry were very much in evidence on the morning of March 28, 1979. A standard cost-minimizing shortcut that the company had adopted was a program for carrying out much of the required routine testing and maintenance on the plant while it was in operation rather than when the reactor was safely shut down. No one would think of performing major equipment checks or maintenance on an airplane in flight—disassembling and cleaning its landing gear, for example—but utility companies commonly do something very much like that to avoid frequent, costly shutdowns of their nuclear units. This effort to keep plants running during testing and maintenance has at least three manifest risks. First, such work might inadvertently interfere with the equipment keeping the plant in steady operation, and so cause a serious accident. Second, standby safety equipment that is deliberately taken out of service for tests or maintenance would not be available (or its operation might be delayed) if an accident that necessitated the use of this equipment occurred while the work was under way. (There are some requirements aimed at ensuring that only a limited amount of a plant's basic safety apparatus will be intentionally disabled at any one time, but these requirements have been neither comprehensively set out nor diligently followed.) A third risk posed by maintenance and testing, regardless of whether it is performed while the plant is operating or shut down, is the possible failure to return equipment to service or to restore the plant to its normal condition after these tasks are completed—for example, a failure to reopen valves that were closed or to restore electrical power that was shut off. The first of these three risks materialized as the immediate cause of the Three Mile Island accident (the maintenance being performed on the polishers, which led to the failure of the main feedwater system), and the third category of maintenance-related safety risks was evidenced in the two closed valves (apparently shut during emergency-feedwater-pump tests two days before the accident, and not reopened) that disabled the entire emergency feedwater system.

* * *

The maintenance that had to be performed on the polishers at Unit 2 was a necessary, unglamorous chore—one of the numberless small tasks connected with the day-to-day operation of the plant. Each of the eight polishers held a twenty-eight-day supply of the resin beads used to filter out impurities from the cooling water. The system had been designed so that only seven polishers would be needed while the plant was operating, and one could be out of service for maintenance. But even though it was intended that maintenance on the polishers would take place when the reactor was running at full power, the safety problems that might arise from this practice were not carefully investigated when the plant was designed. What now appears to be a serious oversight was officially permitted under N.R.C. regulations, because the polishers, like much of the other equipment in Unit 2—including such items as the main feedwater system, the relief valve, and the pressurizer-level instruments, all of which figured so prominently in the March 28 accident—were classified as "non-safety-related" components. Consequently, there were no detailed federal safety requirements governing the design of the polishers, there were no routine federal safety reviews of them during plant licensing, and their performance was not examined during the N.R.C.'s standard plant inspections. Since the polishers were exempted from federal safety rules, they were bought from an outside manufacturer, L.A. Water Treatment, and were installed in Unit 2 without the standard scrutiny that Metropolitan Edison's designers gave to safety-related equipment. Indeed, when investigators went to Unit 2 after the accident they found that the plant officials did not even have accurate engineering drawings and schematics for the Unit 2 polisher system; the drawings that were available showed valves in the wrong places, identified components improperly, and had air-line positions and interconnections incorrectly displayed. Also missing were other types of useful drawings—showing, for example, how the polishers related to other plant systems and equipment.

The periodic removal of spent resins from the polishers had been a chronic problem at Unit 2. The detritus inside the polishers had to be flushed into a receiving tank through a small transfer line that frequently clogged up. Maintenance logs indicate that there was a problem with about one of every twelve transfers. The recurrent difficulties had been reported to the plant's management, but satisfactory corrective action—such as installing alternative filtration processes or modifying the transfer pipe—had not been undertaken. Maintenance workers were simply instructed to try to overcome the blockage by injecting water and compressed air into the transfer line—a seemingly straightforward procedure that actually had serious potential complications, because of leaky valves in the compressed-air system, which sometimes allowed water to seep into it. Since other equipment in the plant—such as instruments that controlled valves directing the flow of water to the main feedwater system—depended on this compressed-air supply, water in the air lines could precipitate a series of malfunctions in the plant. Hence, the problem of a clogged transfer line in the polisher could escalate into real trouble—a fact that had been known for at least seventeen months before the accident at Unit 2.

On October 19, 1977, a maintenance worker removing spent resins from Polisher No. 2 noticed that some of the air-operated equipment on the polisher's control panel had water running out of it. The technician had been using the standard water-plus-compressed-air method of flushing out the polisher's transfer line. Then, suddenly, a set of valves on the polisher system unexpectedly closed, cutting off the flow of cooling water through the polishers and causing the main feedwater system to fail completely. Fortunately, the Unit 2 reactor was not running, since Unit 2 was still in its preoperational testing phase. But the serious implications of this maintenance-related failure were clear. As the event was summarized in an internal memorandum dated November 14, 1977, by John Brummer and Michael Ross, two senior members of the plant's technical staff, the maintenance work had inadvertently created "a total loss of feedwater," so that if the reactor had been operating, "the unit would have been placed in a severe transient"—abnormal—"condition." The memo recommended nine corrective actions, but they were rejected three days later in a cryptic memo from R. J. Toole, the director of G.P.U.'s startup operations; he wrote, "No further action required."

The problem did not go away, however. On May 12, 1978, while spent resins were being transferred out of Polisher No. 7, key valves on the polisher once again unexpectedly slammed shut, disabling the entire feedwater system. Since the reactor was not running at the time, the problem was once again passed over by plant management—this time despite a sharply worded memorandum from the shift supervisor, William Zewe, dated May 15, 1978. Zewe's memo, concerning water in the air and instrument lines, said, "It's time to really do something on this problem before a very serious accident occurs. If the polishers take themselves offline at any high level of power, resultant damage could be very significant." Zewe recommended, as Brummer and Ross had after the earlier loss-of-feedwater incident, that a "bypass" system be installed. If the cooling-water flow *through* the polishers was accidentally cut off by the maintenance crew, a fast-acting automatic bypass that kept the water flowing *around* the polishers could prevent an abrupt disruption of the normal reactor cooling process, all three plant officials said. This recommendation was never acted upon.

Unit 2, in sum, had been operating with a documented history of maintenance-caused failures in its main feedwater system—a situation that had been reported to, but tolerated by, plant management. The accident that occurred on March 28, 1979, initiated by routine work on the polisher system, was in essential respects a replay of the episodes of October 19, 1977, and May 12, 1978—with, of course, the major difference that the Unit 2 reactor, instead of being safely shut down, was operating at ninety-seven percent of full power.

* * *

The types of human error that caused and then exacerbated the Three Mile Island accident were not freak occurrences that can be dismissed as the bad luck of a particular maintenance crew at one individual power plant on a randomly selected day. Government records show that maintenance-caused accidents—and safety-system failures attributable to improper maintenance procedures—are among the dominant, widespread, and recurring safety problems in the American commercial nuclear-power industry. These and other forms of human errors were listed as the cause of eighteen percent of all the serious nuclear-plant malfunctions reported to the N.R.C. in 1978, according to the agency's statistics. To see the Three Mile Island accident as part of this persistent pattern of nuclear-plant safety deficiencies is to understand more fully not only how the accident came about but also what the accident implies about the general state of safety in the other nuclear plants now operating in the United States.

To find documentation of how human error can cause plant accidents, one need dig back no further in the government records than the March 22, 1975, accident that until the events at Three Mile Island had been regarded as the most serious in the nuclear program's history. This accident, which occurred at the Browns Ferry Nuclear Plant, near Decatur, Alabama, involved a fire in the plant's electrical system that was started by a workman's candle. The plant, which, like Three Mile Island, had been in commercial operation for only a few months, was undergoing a postconstruction modification, and this work required technicians to check for air leaks in a section of the electrical system where some cables passed through the reactor-building wall. The plant had three nuclear reactors, two of which were operating at full power. An electrician's aide, twenty years old and untrained, who had been on the job only two days, was holding a lighted candle and watching its flame flicker as an indication of possible air leaks. He found a leak, and in the process accidentally set a fire that burned uncontrolled in Unit 1 for seven and a half hours and badly damaged sixteen hundred electrical cables, including six hundred and eighteen cables that were connected to plant safety systems. The fire burned through the cables and swiftly crippled the plant—and especially its emergency cooling apparatus. The error resulted in a common-mode failure that simultaneously incapacitated multiple safety systems. The plant's superintendent, Harry J. Green, has commented, "We had lost redundant components that we didn't think you could lose." Seven years later, the debate about how narrowly the Browns Ferry plant escaped a meltdown accident is still going on.

Another accident related to improper testing and maintenance occurred at the Zion Nuclear Power Station, in Zion, Illinois, on July 12, 1977. Because of what Dr. Hanauer referred to in an N.R.C. internal memorandum, dated August 18, 1977, as an "obvious gross management deficiency," worsened by an "unsafe" Westinghouse design for the plant's pressurized-water reactor, the plant experienced a serious accident. What happened, quite simply, was that technicians, in the process of checking some of the safety-system circuitry while preparations were being made for restarting the plant after a brief shutdown, fed thirty-one "dummy" instrument signals to the control room that made it appear that the reactor had an adequate cooling-water supply when, in fact, it was losing water through a drain line. The plant safety systems, which would have automatically supplied the reactor with more water under normal circumstances, failed because they, too, received the false instrument signals. In other words, the safety equipment that had been provided to control this accident was simultaneously paralyzed—as a result, once again, of a common-mode failure. Forty minutes later, after several thousand gallons of cooling water had been drained out of the reactor, the operators finally recognized the problem—fortunately, in time to correct it before the water supply had declined to critically low levels. Dr. Hanauer's memorandum warned that "next time, some different and not now foreseen sequence of events may start the ball rolling," and urged a broad review of all plants, "Westinghouse and non-Westinghouse," to prevent the recurrence of such an incident. The review was never carried out.

Despite these examples and others, and despite the estimates from the N.R.C.'s own eight-volume *Reactor Safety Study* that testing and maintenance were among the leading expected causes of safety-system "unavailability" during reactor accidents—thirty-five percent of the failures in reactor-protection systems would be caused by testing or maintenance, according to this study, which was done in 1975—no satisfactory remedial action has been taken by the N.R.C. Unlicensed technicians are still allowed to carry out work on sensitive plant components. No regulations have been issued that require plants to be modified systematically so that their equipment can be tested or maintained in ways that would reduce the likelihood of accidents.

Thus, while great care is supposed to be taken in designing the essential equipment of nuclear plants, testing and maintaining this equipment have been left a largely unregulated aspect of plant operation. Moreover, despite federal requirements that each plant be provided with a broad assortment of sophisticated safety devices, federal nuclear-safety authorities sanction industry practices allowing this equipment to be intentionally disabled for tests and maintenance during plant operation. The official listing of each plant's safety systems is, therefore, something like a menu that might be offered by a gifted but erratic chef: it has impressive entries, but what is available on any given day—such as March 28, 1979—may be just potluck.

* * *

A number of measures, of varying degrees of cost and effectiveness, can be taken to prevent workers from meddling with equipment, disabling safety systems, and otherwise interfering with the safe operation of a nuclear-power plant. The preferable approach to the problem of human error is for plant designers to provide positive physical safeguards aimed at protecting all critical equipment in the plant from inadvertent or inappropriate human tampering. Such measures could be compared to some of the steps that many parents take in an effort to childproof a house—capping electrical outlets, installing special locks on windows, putting gates on staircases. Instead of providing safety features of this type in a nuclear plant, one can, of course, simply tell the nuclear-plant technician, like the child, not to do certain hazardous things. While specific, safety-related instructions, careful training, and repeated exhortations would be of considerable value, there is little doubt about the superiority of physical safeguards.

Comprehensive automation is obviously the most far-reaching countermeasure that can be taken at a nuclear power plant to reduce the opportunity for human error. The specific safety procedures stipulated by the designers can be locked into the electronic memories of the plant's control devices, which will then issue the commands that direct all critical operations. Automated control systems can be exhaustively tested in an effort to ensure that they unfailingly adhere to the designers' orders. Though no system would be flawless, the technology for automated control equipment has progressed to the point where the machine is vastly more reliable than human beings—who are sometimes capricious, forgetful, distracted, tired, confused, error-prone, or negligent. Unfortunately, the automated systems required to carry out all the complex tasks associated with nuclear-plant operation are expensive, so American nuclear-utility companies are reluctant to use them except for a limited set of plant functions. The companies prefer to rely on relatively low-salaried and often unskilled technicians to perform what they regard as mundane chores. Even under emergency conditions, the automated equipment that is in general use has a narrow function: it is relied upon merely to turn on safety equipment. Once this is done, human operators rather than carefully programmed machines are responsible for all remaining aspects of crisis management. The current division of labor between man and machine during nuclear-power-plant accidents was illustrated quite precisely during the accident at Three Mile Island: two minutes after the accident began, automated command signals turned the plant's emergency core-cooling system on; the control-room operators then took command and, some two minutes later, decided to turn this system off.

There are steps short of full automation that can be taken to mitigate potential human errors. For example, valves that must be kept open or closed to ensure that the safety system works properly can simply be locked into the correct position before a plant is started up, and access to the keys can be limited. Or else more sophisticated safeguards can be placed on the switches in the control room that plant operators use to open and close the valves. One such safeguard consists of electronic interlocks to link the switches that operate equipment with the instruments that monitor plant conditions; these interlocks can be set up to make sure that the operation of key components and systems is always matched to the safety needs of the reactor. One system of interlocks might be installed to prevent the switches that turn the reactor on from working unless all key valves are in their proper positions. A whole series of such interlocks, imposed not only on valves but also on all other safety-related equipment, would ensure that the reactor could be started only if all its safety equipment is in the desired condition. Another type of interlock could be installed so that once the reactor is operating, safety-related valves and other equipment could not be switched off or otherwise inappropriately manipulated by a mistaken command from the control room. If the control-room operator tries to execute a wrong maneuver or pushes the incorrect button, the control system could automatically cancel this action and prevent a disruption of the plant.

Instead of incorporating into the plants permanent design features to prevent human errors, American nuclear-utility companies—with the permission of the N.R.C.—rely on one of the weakest of all possible systems: volumes of written procedures that simply tell nuclear-plant employees what they should and should not do. The members of the work force at a nuclear plant cannot be expected to remember each step in the formal rules that guide their every safety-related action. Nor, since they are generally required to have only high-school diplomas, can it be hoped that they will always be able to understand the technical rationale behind the instructions given to them. The plant management, accordingly, is supposed to set up a systematic "quality-assurance" program that will supervise the performance of plant workers, will check and double-check to see that procedures are being properly followed, and will maintain careful written records documenting painstaking adherence to all the dos and don'ts that are intended to govern day-to-day nuclear-plant activities. "The watchword throughout the nuclear reactor industry is *quality assurance,*" according to the nuclear pioneer Alvin Weinberg, who directed the A.E.C.'s Oak Ridge National Laboratory from 1955 to 1973. In other industries, the quality of workmanship and the resultant product reliability are sometimes uneven and unpredictable, but the nuclear-power industry claims to have achieved an unprecedented level of meticulousness. This (by prevailing industrial standards) superhuman diligence is, according to the government's official safety philosophy, the principal guarantor of nuclear-power-plant safety.

Although running nuclear-power plants by the book is a major part of the N.R.C.'s approach to safety, the commission, like the A.E.C. before it, takes little official interest in the actual drawing up of the detailed procedures that nuclear-plant workers are supposed to follow. It delegates the task of writing all procedural manuals to the utility companies, asking them to follow what are sometimes only generalized N.R.C. guidelines. Moreover, even after the utility has prepared its procedural manuals—which cover plant startup, shutdown, maintenance, equipment installation, emergency procedures, safety reviews, overall plant quality assurance, and other activities with major safety repercussions—the N.R.C. carries out no comprehensive reviews to determine whether the procedures adopted by the company actually represent sound practice and conform with the N.R.C. guidelines. Federal nuclear-safety regulators, by their own estimates, actually inspect only about one to two percent of the safety-related activities at a plant.

The utility companies themselves are hardly enthusiastic about the responsibilities assigned to them in this area. The task of writing procedural manuals is detailed, time-consuming, often boring and, in the companies' eyes, unproductive; they regard much of the effort as creating useless paperwork that wraps all aspects of a plant in bureaucratic red tape. The net result of the N.R.C.'s dependence on a self-regulating nuclear industry to develop and follow strict safety procedures and of the industry's indifference, even hostility, to these requirements was evident at the Browns Ferry nuclear plant. The investigation of the Browns Ferry fire disclosed that, contrary to a technical specification in the plant's license, no detailed procedures had been drawn up and put in writing to govern the maintenance activities that caused the fire. In addition, according to the N.R.C.'s private files, the government was fully aware that the Browns Ferry plant did not have a competent quality-assurance program. "An overall QA program acceptable to NRC has not yet been produced by TVA"—the Tennessee Valley Authority, which operates the plant—Dr. Hanauer wrote in an internal N.R.C. memorandum dated July 10, 1975. Accident investigators, he noted in this brief memo, which was sent to senior N.R.C. officials, "asked very embarrassing questions about how [Browns Ferry] Units 1 & 2 had operating licenses without acceptable QA programs." Questions of this nature might also be addressed to the N.R.C. about Three Mile Island.

* * *

Just how carefully the safety-related quality-assurance procedures at Three Mile Island Unit 2 had been prepared, how closely they were adhered to by plant personnel, and what role they played in bringing about the accident can be seen from plant records. At ten on the morning of March 26, 1979—two days before the accident—technicians at Three Mile Island Unit 2 began what the written plant procedures described as routine surveillance testing of the plant's emergency feedwater pumps, which called for the technicians to close two "isolation" valves. The technicians wanted to run these pumps to check on their operability, but they did not want to let water from the pumps be discharged into the plant's cooling network and so interfere with reactor operation. The simultaneous closing of these two valves, however, prevented the emergency feedwater pumps from delivering water to the parts of the plant's cooling system where it would be needed during certain accidents. At a meeting that took place seven months before the accident, a review committee made up of Metropolitan Edison employees had approved the closing of the two valves during pump testing. The employees' decision had received no independent review or checking by Nuclear Regulatory Commission inspectors; nor did the plant's review committee itself, according to plant records, review the safety implications of shutting the two valves simultaneously.

If the committee and the N.R.C. had adequately reviewed the procedures for testing the emergency feedwater pumps, they would have found that closing the two valves during plant operation was a direct violation of conditions that are set forth in the technical specifications incorporated in the plant's federal license. The technical specifications require that three emergency feedwater pumps and the piping system connecting them to the rest of the plant be operable at all times when the reactor is going. (The only exception is that one pump at a time may be taken out of service for maintenance provided that it is restored to operable status within seventy-two hours or the plant is shut down within the succeeding twelve hours.) The complete disabling of the emergency feedwater system is not allowed for any purpose during plant operation, yet plant records show that the system was shut down for testing on January 3, 1979, and February 26, 1979, as well as on the morning of March 26, 1979. And although the N.R.C. did add to the plant's license the technical specification intended to ensure the operability of the emergency feedwater system, the N.R.C. required no physical safeguards to be installed at the plant—locks on the two valves or electronic interlocks on their control switches—to make sure that the valves remained open and the emergency feedwater system operable when the reactor was running.

Once the testing of the pumps was completed on March 26—this took about three hours—the two isolation valves should have been reopened, as the test procedures specified. This is an essential step in restoring the emergency feedwater system to operability. Martin Cooper, one of four plant employees who were involved in the test, has said that he reopened the two valves at the completion of the test. "I actually opened the valves myself," he told a presidential commission appointed to investigate the accident. "I was the control-room operator on duty." Cooper has also explained that a checkoff sheet on which the plant staff recorded its adherence to the procedures for reopening the valves was "thrown in the trash can," and that only the data sheets—the record of how the pumps performed during the testing—were kept in plant files. Moreover, no plant-management personnel reviewed either the actual completion of the procedures or the checkoff sheets; senior officials merely looked at the data on pump performance, and did not certify that the emergency feedwater system had been satisfactorily returned to service.

Both the failure to retain the checkoff sheets and the failure of plant management to oversee the testing are in direct conflict with two fundamental plant-safety procedures. These require that supervisory personnel review not only the results but also the documentation of test procedures, and that records of surveillance of test procedures be retained for at least five years—requirements that keep these activities under careful management control and serve as a check on the management's own performance. The failure to adhere to these two procedures, it has been discovered, was not limited to the particular testing procedure on March 26, 1979. Plant-management personnel have admitted to N.R.C. investigators that, as a de facto rule, completed test procedures at Three Mile Island Unit 2 were hardly ever reviewed because of the length of the procedures and the burden of the general management workload. They have also said that the required records documenting the plant testing activities were generally not retained because of a shortage of storage space.

In sum, the testing of the emergency feedwater pumps on March 26 should not have been done in the first place, since the basic testing procedure violated explicit federal safety restrictions imposed on the plant and, in addition, was carried out in contravention of general plant procedures themselves of considerable importance to plant safety. Despite this serious breakdown in plant safety precautions, the testing personnel still assert that the valves whose closure disabled the emergency feedwater system on March 28 were reopened at the completion of the testing two days earlier. To date, however, there is no evidence available beyond the say-so of these employees that this was in fact done. Moreover, plant records show no other maintenance or tests of any type being performed on the emergency feedwater system, in the period between the testing on March 26 and the accident, that could have resulted in the closure of the valves. The many investigations of the accident have brought forth no new evidence, so the most probable cause for the closure of the two valves appears to be a mistake by the testing crew on March 26.

* * *

A valve controls the flow of a liquid or a gas. Some valves in a nuclear power plant are no more complicated in design or function than the faucet on a sink, while others, designed to control the flow of thousands of gallons of water per hour through a nuclear plant's main cooling-water arteries, are complex giants that weigh many tons, stand several feet high, are opened and closed by powerful motors, and must be precisely controlled and monitored by electrical systems. Such valves must meet a variety of complicated design standards and receive periodic testing and maintenance.

Many of the valves perform tasks other than the regulation of cooling-water flow. Some feed fuel to diesel generators that provide an emergency power supply for plant safety systems. Some control the flow of compressed air, which is used to perform a variety of functions; indeed, certain valves may themselves be opened and closed by compressed air, and thus their proper functioning depends on other valves. The containment building of a nuclear plant is supposed to be sealed in the event of an accident, so that radioactive materials will not leak out, and to this end a variety of containment-isolation valves are installed, which are designed to shut off nonessential piping leading into and out of the containment building at the first sign of a serious accident. In some containment buildings, large valves are routinely open so that low-level radioactivity, excess air pressure, and heat can be continuously vented to the outside, and, obviously, these so-called containment-purge valves must close quickly in the event of an accident. Certain valves in nuclear plants relieve excessive pressure in the reactor. Some of these valves open automatically when the pressure of the cooling water inside the reactor increases beyond a set point. Others can be opened or closed at the discretion of the operators.

Although valves are an integral component of all major nuclear-plant safety systems, plant owners and operators have not taken proper precautions to ensure their reliable performance. As a consequence, the general performance of the thousands of valves installed in each nuclear-power plant is very much below par—a fact that has been evident for years. According to a short internal memorandum written in 1972 by Dr. Hanauer—his colleagues sometimes refer to his missives as Hanauergrams—nuclear-plant valves "are still failing too often and for the same reasons over and over." A longer report by Hanauer the same year noted:

> Valves do not have a very good reliability record. Recently, five of the vacuum relief valves . . . of Quad Cities 2 were found stuck partly open. Moreover, these valves had been modified to include redundant "valve-closed" position indicators and testing devices, because of recent [Atomic Energy Commission] concerns. The redundant position indicators [devices to alert the control-room operators to the fact that the valves were stuck open] were found not to indicate correctly the particular partly open situation that obtained on the five failed valves.

If the valves were in the wrong position, Dr. Hanauer's report went on, the containment chamber could overpressurize and rupture under some accident circumstances, whereas under other accident circumstances the improper operation of these valves could cause the pressure in the containment building to fall so low that the structure might collapse.

Despite these warnings, no systematic remedial action to

improve the performance of valves in nuclear plants has been taken in the years since 1972. (Dr. Hanauer's particular recommendations for dealing with the problem affecting Quad Cities and dozens of similar plants were rejected by Joseph Hendrie, the N.R.C. chairman, who was then a senior A.E.C. official. Hendrie wrote, in a memo dated September 25, 1972, that Hanauer's recommendations would reverse "hallowed policy" and would be so embarrassing for the A.E.C. that such a reversal "could well be the end of nuclear power.") N.R.C. records show that valves have remained a major problem for nuclear plants.

The nature and the extent of the continuing valve problems have been well documented by Dr. Hanauer in a special N.R.C. internal file that he personally set up on accidents and safety defects at American nuclear-power plants. A man who is able to view the blemishes that compromise nuclear-plant safety with a degree of dry humor, Dr. Hanauer calls this record the Nugget File. Some of the malfunctions documented in the Nugget File involve serious accidents, but the majority of the failures recorded in the file had few or no direct safety consequences. Since the safety apparatus involved was mostly standby equipment—normally inoperative but called into service when a certain type of accident arises—a failure of this equipment that took place during periodic testing or was uncovered in routine inspections obviously involved no immediate threat of the public's being exposed to radiation. Such failures are not instances of near-catastrophes; they merely indicate just how unready the nuclear plants are to cope with accidents when they arise. The Nugget File is chiefly a catalogue of these warning signs—case studies giving evidence or intimations of weak spots in the safety precautions taken at American nuclear-power plants:

*September 1968:* An operator at Unit 1 of the Dresden Nuclear Power Station, in Morris, Illinois, noticed that the monitoring lights for motor-operated valves in certain cooling systems were not on—indicating a loss of power to the mechanism that opens and closes the valves. Electricians determined that the circuit breakers for these valves were not working, for the reason that an electrical-distribution panel had water leaking into it. During a heavy rainfall, water had accumulated on the roof of the plant, because of a plugged drain, and had subsequently seeped down a building column and into the electrical system controlling the valves.

*April 1969:* During the shutdown of a reactor, the name of which was not specified, the heat-removal system did not supply cooling water to the fuel core for two hours. Earlier, while the pump was running, a technician performing routine maintenance had closed a valve supplying water to the decay-heat pump, even though a special maintenance procedure prohibited the closing of this valve without the consent of the control-room supervisor.

*April 1971:* Six months after the Point Beach Nuclear Plant, in Two Rivers, Wisconsin, received its operating license, a test revealed that two key valves in the emergency core-cooling system would not open when they were signaled to open. Both valves were designed to open and close by hydraulic pumps, but the pumps had been installed incorrectly: they were mounted horizontally rather than vertically. On a prior occasion, a short circuit had disabled the hydraulic system, and on another occasion a valve in the hydraulic system that was supposed to open had stuck shut. Neither the utility company's quality-assurance program during the construction of the plant nor an investigation of the two earlier valve failures had disclosed the incorrect installation of the hydraulic pumps.

*June 1973:* While the H. B. Robinson Steam Electric Plant, in Hartsville, South Carolina, was operating at seventy-five percent of capacity, the plant safety committee determined that certain instruments were incorrectly measuring the pressure in the reactor building. The purpose of these instruments was to turn safety systems on in the event of a high-pressure reading, which would be indicative of a pipe rupture in the primary cooling system. The pressure-sensing instruments had failed because three vent valves on them had been left open. The three valves were not on the valve checklist used by plant operators, and were not even on the plant-equipment drawings.

*October 1973:* The spray systems for the containment building at the Oconee Nuclear Station Unit 2, in Seneca, South Carolina, were found to be inoperable; these spray systems were intended to prevent the rupture of the containment building during an accident. They were inoperable because their valves were inadvertently left closed after a test that had been conducted eleven days earlier. The company that owns the plant said that as a "corrective action" it would issue a memorandum to the plant staff emphasizing the need for close attention to procedures.

*May 1974:* The Browns Ferry Nuclear Plant was notified by the T.V.A. design division that improper valves had been installed in the primary cooling system of the plant's three reactors. The valves were designed to operate at a pressure of six hundred pounds per square inch, but had been installed in a system with a pressure more than twice that. The design division stated that a hundred valves of this inappropriate design had been installed in each of the three reactors; they had been in operation in Unit 1 for almost a year.

*June 1975:* The emergency-core-cooling systems at the Maine Yankee Atomic Power Plant, in Wiscasset, Maine, were found to be inoperative because valves were in the wrong position. The system was designed with a fail-safe valve arrangement that would automatically discharge emergency cooling water into the reactor in the event of a major pipe rupture. The fail-safe design was negated when the discharge valves for all the emergency cooling-water tanks were found closed. The operator who completed the system's valve checkoff sheet noted the locked handwheels on these valves and mistakenly assumed that the valves were locked open when in fact they were locked closed.

*September 1975:* At Unit 2 of the Oconee Nuclear Plant, the level of reactor-cooling water was accidentally altered while the reactor was operating at full power. Workers trying to shut off valves so that they could make repairs to a faucet in the plant's chemistry laboratory inadvertently closed valves controlling the air supply to two other valves. These other valves controlled the rate of water drainage from and addition to the reactor's primary cooling-water system. Before this error, the two flows balanced each other, but after the air supply was closed, one valve failed in a closed position, preventing drainage, and the other valve failed in an open position, allowing an excess addition of cooling water.

*December 1975:* A plant operator at the Calvert Cliffs Nuclear Power Plant Unit 1, near Lusby, Maryland, discovered that a water-supply valve to the two emergency feedwater pumps was shut, leaving both pumps without a supply of water. The Baltimore Gas & Electric Company, which owned the plant, concluded that plant operators had "erred" about two weeks earlier when the valve positions were changed. The company report on this "occurrence" said that if the emergency feedwater pumps had been needed, it was "highly probable" that the operator would have noticed the lack of water supply to the pumps "prior to any serious damage occurring."

*March 1976:* At the Prairie Island Nuclear Generating Plant, in Welch, Minnesota, a valve was wrongly closed during maintenance on the cooling-water system for plant safety apparatus. This single valve in the wrong position disabled one of the two cooling systems for the two reactors at the plant.

*July 1976:* A steam-driven emergency feedwater pump at Unit 2 of the Millstone Nuclear Power Station, in Waterford, Connecticut, failed because the steam-supply valve would not open. An inspection showed that a steel piece of the operating mechanism had broken off. A subsequent investigation revealed multiple contributory failures. The switch that was used to control the motor that opened and closed the valve had failed. A backup switch had also failed. In addition, the valve had not been tested; it had initially been installed as hand-operated equipment, and had been returned to the manufacturer so that a motor could be put on to open and close it. The valve had been reinstalled without an adequate inspection or test.

Dr. Hanauer is still enlarging the Nugget File. "Yes, sir, it's still an open file. I just put an item in the other day," he said in a recent interview. Valves, he added, were "one of the continuing problems." Another item that deserves to be included in his collection of valve-related problems occurred at the Arkansas Nuclear One Station—Unit 1, in Russellville, which uses a Babcock & Wilcox reactor. During routine testing of valves on the plant's main feedwater system, technicians mistakenly disabled the automatic controls for the plant's emergency feedwater system—the same system that was rendered inoperative at Three Mile Island Unit 2 on March 28, 1979. In both instances, improper testing procedures were used. The Arkansas Unit 1 problem occurred on June 2, 1979—more than two months after the accident at Three Mile Island.

It is evident from the Nugget File that valve failures, many resulting from improper maintenance and sloppy operating procedures, are endemic to the commercial nuclear-power industry in this country. As disturbing as this finding may be, an even more sobering conclusion emerges if one reads through the several hundred technical reports that are included in the file. What these reports reveal is that the same level of carelessness and management quality-assurance breakdowns that shows up in the records of valve failures apparently also infects every other major aspect of the design, construction, and operation of the seventy-one nuclear plants now in operation around the country. For example, it is not just maintenance on valves that leaves safety apparatus disabled. Electrical-system maintenance, which generally necessitates unplugging equipment or otherwise interrupting power-supply circuits, has a record parallel to that of valve maintenance in leaving safety apparatus inoperable; just as workers forget to reopen valves, they commonly forget to turn power supplies back on or to plug safety devices back in. Moreover, these blunders extend far beyond maintenance and testing activities; they persistently occur in the design of equipment, in plant construction, in equipment installation, and at the operating consoles in control rooms. Thus, according to the Nugget File, key safety equipment is often rendered inoperative for want of fuses. Electrical relays fail because they are painted over or welded together or disconnected—or simply left out when the equipment is installed. Nuclear-reactor control rods fail to work because their components are installed upside down or because their electronic-control systems pull them out of the reactor when the operator presses the button to put them in. Sensitive pieces of safety apparatus malfunction because they are frozen or burned or flooded or dirty or corroded, or have been bumped or dropped or overpressurized or unhinged or miscalibrated or miswired. In some cases, they are, to cite one of Dr. Hanauer's marginal notes in the Nugget File, "guaranteed not to work" because of bad initial design. This possibility was displayed quite dramatically in July 1980, when the SCRAM system—a supposedly ultra-high-reliability safety feature that acts as the "emergency brake" that instantly shuts off the reactor—failed at the Browns Ferry Nuclear Power Station Unit 3 in Alabama. The system failed repeatedly as the operators, over a period of fifteen minutes, kept trying to insert the control rods into the reactor. One industry newsletter, reporting the dismay in the industry over the failure of such a basic safety system, described the problem as "mysterious." Actually, what was involved was a simple design error that had been overlooked by federal safety authorities, despite the fact that the SCRAM system had been relied upon for two decades in some twenty-six plants. There have been other kinds of problems with SCRAM systems over the years, not all of which involved a failure to shut down the reactor. In some cases, the problems have involved the unintentional Scramming of the reactor when it should have been left operating. In one instance at a training reactor, the SCRAM system was activated because of a change in the water pressure in the reactor building—which happened when someone flushed a toilet.

Such failures are manifestations of the pervasive breakdown of the "honor system" that the N.R.C. has set up—the loose arrangement under which utility companies are expected to police the safety of their own nuclear plants. This honor system appears to have resulted, in an uncomfortably large number of cases, in little discipline being exercised over a wide range of plant activities important to public safety. That the relief valve in the primary cooling system and two key valves in the emergency feedwater system of Three

Mile Island Nuclear Station Unit 2 should happen to be in the wrong position on the morning of March 28, 1979, is no more surprising than it would have been if the company that operates the facility, Metropolitan Edison, had decided that day to apply for a rate increase to cover rising costs: such events occur in the nuclear-power business with what seems to be roughly the same frequency. The repeated instances of the disabling of sophisticated safety systems by crude human errors have led to the view, expressed by Henry Kendall, a physicist at the Massachusetts Institute of Technology, that the nuclear-power program may deserve to be described, like the Navy in *The Caine Mutiny,* as a system designed by geniuses and run by idiots.

# 2: The Paper Trail

The safety of the commercial nuclear-power plants in this country is the responsibility of the United States Nuclear Regulatory Commission, a small federal agency whose overall performance—and basic technical and administrative competence—has been widely criticized in light of the accident at Three Mile Island. In addition to the presidential commission set up to investigate the accident, the N.R.C. undertook its own postaccident self-examination, conducted by a Special Inquiry Group headed by the Washington attorney Mitchell Rogovin, which passed a strongly negative judgment on the N.R.C.'s regulatory program. "In our opinion the [N.R.C.] is incapable, in its present configuration, of managing a comprehensive national safety program for existing nuclear power plants and those scheduled to come online in the next few years adequate to insure the public health and safety," it concluded.

In one sense, sweeping general criticisms of the N.R.C.'s regulatory program may be unfair, for they may be belaboring "deficiencies" in its performance that are not so much the result of sheer regulatory ineptitude as they are the consequence of political decisions about what the agency was expected to do in the first place. The fact is that the N.R.C. was never really intended to tightly constrain the nuclear-power program or exercise ultimate judgment about the acceptability of the risks that might be associated with it. Given the overriding commitment of the federal government to rapid nuclear-power expansion—a bipartisan national goal adopted by Congress in the 1950s and endorsed by successive administrations—the N.R.C.'s role in the effort was a largely *pro forma* one. Despite what legislators and bureaucrats refer to as the "boiler plate" language in the statutes that define the N.R.C.'s regulatory powers and duties, the agency, in practice, has been simply the minor government department that hands out licenses to build and operate nuclear plants, much as the State Department's Passport Agency issues passports to travelers. The safety and licensing system set up by the N.R.C., according to Peter Bradford, one of the agency's five commissioners, "was never designed to be an effective regulatory system" but "was constructed for the stamping out of nuclear power plant construction permits or operating licenses at the assembly-line rate of one or two per week for every week in the years from 1975 to 2000." One consequence of the government's single-minded dedication to nuclear-power expansion, Bradford notes, was a regulatory process that fell into "fundamental disarray."

The regulatory arrangements relied on by the N.R.C. are the outgrowth of general policies adopted in the 1950s and 1960s by its predecessor, the Atomic Energy Commission. The A.E.C., which vigorously promoted commercial exploitation of "the peaceful atom," believed that the shortest route to a large nuclear-power program required unleashing "the genius and enterprise of American industry." In order to give the nuclear-power industry the discretion it needed to move ahead swiftly with plant construction, the A.E.C. adopted a relaxed and permissive regulatory program. Under the scheme it developed, the government would specify general safety goals but would leave their implementation to the emerging nuclear industry, relying on it to have the common sense to build a high level of safety into each plant. The A.E.C. did have a small number of staff engineers and consultants who reviewed the applications for nuclear-plant licenses, but the commission usually gave short shrift to the safety issues raised by its experts. Since the A.E.C. was convinced that technical solutions to these problems could be found, it did not believe that it was running a great risk by deferring such questions while the industry proceeded with nuclear-plant construction. Moreover, when the industry matured and expanded, the A.E.C. still saw no strong need to develop more formal regulatory arrangements. In 1974, however, Congress decided that the conflict of interest inherent in the A.E.C.'s dual responsibilities—to both promote and regulate the nuclear industry—suggested the need for a new, independent regulatory body. Accordingly, Congress passed the Energy Reorganization Act of 1974 to establish the N.R.C.—a separate agency whose only mission would be to regulate the nuclear industry. Reorganization per se did not mean substantive regulatory reform, however, for there was little Congress could do about the bureaucratic inertia that would inevitably make it difficult to change the ground rules for nuclear-power expansion laid down by the A.E.C. Indeed, the carry-over of A.E.C. policies by the N.R.C. was a foregone conclusion, since the N.R.C., according to the 1974 act, would essentially *be* the regulatory staff of the old A.E.C., under a new name. Unsurprisingly, the new agency's first official action was to adopt all the safety rules and policies of its predecessor. As a matter of law, the N.R.C. had full and unambiguous power to develop new regulations governing every aspect of nuclear-plant design and construction, all the way down to rules for the installation of the last nut and bolt. Still, the only basic changes in the regulatory process during the next several years were efforts—prompted by the Ford and Carter administrations—to streamline it so that nuclear plants could be licensed faster.

With a regulatory process that remained in "fundamental disarray," it was apparent—to some observers, at least—that sooner or later the government's overambitious nuclear program was going to get into trouble. Indeed, looking back over A.E.C. and N.R.C. records, one can follow what one senior N.R.C. official calls the paper trail that documents the detailed foreknowledge, on the part of both the nuclear industry and the federal government, of the specific safety problems that culminated in the accident of March 28, 1979.

* * *

Three Mile Island Unit 2 is one of nine electric-generating plants in the United States that use nuclear-reactor equipment designed and manufactured by Babcock & Wilcox. As is customarily the case with nuclear-equipment suppliers, Babcock & Wilcox provides only the specialized nuclear steam-supply system and the associated control equipment used in these facilities; its portion is estimated to amount to about ten percent of the total cost of plant construction, with other suppliers, architect-engineering firms, and construction companies providing all the other services and equipment. (Unit 2, for example, was built by United Engineers & Constructors; its turbine was supplied by Westinghouse; its architect-engineer was Burns and Roe; the pressure-relief valve for its primary-cooling system was manufactured by Dresser Industries.) The various companies that participate in the design and manufacture of each nuclear-power plant's components and systems have their own design preferences and requirements, and so do the individual utility companies buying these services; as a result, the nine plants that use Babcock & Wilcox nuclear reactors differ in many of their design details. The nine plants were also built at different times and were therefore subject to different federal safety regulations. Even plants built during the same period and under the same nominal safety requirements may display marked design differences, owing to the varying interpretations that can be given to some of the very general design criteria promulgated by the A.E.C. and the N.R.C.

Still, the plants using Babcock & Wilcox reactors do have a strong family resemblance. Certainly the Babcock & Wilcox equipment in all of them is essentially the same in most elementary design respects. The official A.E.C. "Safety Evaluation Report" for Three Mile Island Unit 2, issued in September 1969, noted, "The nuclear steam supply system, engineered safety features, and reactor building are similar in design to the other Babcock & Wilcox facilities at Oconee Nuclear Station, Crystal River Nuclear Station, Rancho Seco Nuclear Station, Three Mile Island Unit No. 1 Nuclear Station, and the Arkansas Nuclear Station, which have already been issued construction permits."

The eight other nuclear plants incorporating Babcock & Wilcox reactors all went into operation before Three Mile Island Unit 2, and as of March 1979, the nine plants had accumulated some thirty reactor-years of operating experience. The three Oconee plants, which are owned by the Duke Power Company and are situated near Seneca, South Carolina, went into operation in 1973 and 1974. Toledo Edison's Davis-Besse Unit 1, at Oak Harbor, Ohio, between Toledo and Cleveland, has been operating since 1977; the Florida Power Corporation's Crystal River Unit 3, near Crystal River, Florida, since 1976; the Sacramento Municipal Utility District's Rancho Seco Nuclear Generating Station, in Sacramento, California, since 1974; the Arkansas Power & Light Company's Arkansas Unit 1, in Russellville, Arkansas, since 1974; and Three Mile Island Unit 1, which, like Unit 2, is operated by the Metropolitan Edison Company, since 1974.

The operating records of the earlier Babcock & Wilcox nuclear reactors provide concrete, practical information about the performance characteristics and overall reliability of Babcock & Wilcox systems and components. The design of nuclear-power plants is based on predictions of how equipment will behave under both normal and emergency conditions. While laboratory tests and experiments can provide some check on the design engineers' predictions, operating experience affords far more realistic and valuable data. Thus, as nuclear-power plants of a given type and design operate, careful study of their cumulative track record can serve a number of ends. The operating history can be used to check design adequacy; it can suggest needed procedural improvements; it can prompt new regulatory requirements or, possibly, the elimination of unnecessary ones; it can be used as a basis for other regulatory determinations (such as the competence of individual utility companies to operate nuclear-power plants); and, most important, it can provide possible early-warning signs of developing safety problems.

The N.R.C. has never given high priority to an effort to study and use the operating records of commercial nuclear plants as a basis for its regulatory actions. Its scientists and engineers are otherwise engaged. N.R.C. safety analysts, for example, perform prelicensing safety reviews of each proposed nuclear-power plant; they are concerned principally with the general design features of new plants and, as a rule, limit their inquiries to standard questions about safety-system design that have been on the official technical agenda since the mid-1960s. Meanwhile, in the various government laboratories where N.R.C. safety research is conducted, scientists and engineers concentrate on a relatively narrow set of technical questions about certain "hypothetical accidents"—advanced scientific issues that have been the dominant interest of these research centers for the last fifteen years. Thus, the federal agency in charge of nuclear-plant safety, given the duties and preoccupations of its technical experts, has concerned itself only casually with the actual day-to-day safety problems at American nuclear plants.

The N.R.C. does have a Division of Operating Reactors and an Office of Inspection and Enforcement, and these presumably are responsible for reviewing the safety problems of nuclear plants already in service. According to Victor Stello, Jr., who was the head of the former division and is now the head of the latter, the N.R.C. "collects, digests, analyzes," and otherwise deals with the reports from the operating plants, but just how these efforts fit together into an overall system Stello was reluctant to explain. "No, I don't like to summarize it," he said in an interview a few weeks after the accident. The task of reviewing these data, he remarked, is spread among various agency divisions and offices. "There's no assigned group that sits down and is devoted to the task of just looking at all of the licensee event reports from all of the plants—it's not done that way," he said. Some offices do collect the data from the plants, he continued, adding, "I would think" that they "do what they can to analyze it." Other N.R.C. branches, according to Stello, have "a whole host of people" who analyze individual events at the operating plants "fairly comprehensively."

Harold Denton, who as the director of the Office of Nuclear Reactor Regulation supervises these specialized branches, has a different impression of how the N.R.C. safety analysts have dealt with routine reports of equipment malfunctions and other safety problems over the years. The reports "all come in, they all get logged, and identified, and printed up," Denton says, but "some branches are far more interested in operating experience than others." While a few offices have "historically really monitored" and "worried" about such developments at the licensed reactors, he says, "other branches don't tend to think that operating experience is as valuable in input, and it hasn't got done." After the Three Mile Island accident, Denton said, "The commission has asked us on a crash basis now to propose a new scheme to be sure that licensee event reports do get proper attention in the future."

Even Stello conceded that the N.R.C., before the accident, had no coordinated method for identifying patterns in the safety-system deficiencies reported to it. "The individual reports are analyzed," he said. "The trending is not. The trending is ad hoc."

Unless some N.R.C. officials merely happened to discover it, a trend suggesting safety difficulties could easily remain buried in the voluminous reports that the N.R.C. receives from each operating plant. Over the years Hanauer has sent out memos on the need to review the information flowing into the agency from the plants. "Not a day goes by without one or more mishaps at licensed reactors," he noted in a memo on September 13, 1971, in which he urged the orderly compilation and "periodic review" of these data. Other officials have privately discussed the "gold mine" of safety-related data that were being officially ignored, although a few abortive attempts were made to set up a program for extracting useful—and possibly critically important—information from all the routine reports of nuclear-plant mishaps. Still, the N.R.C.'s safety-review and safety-research priorities had never been amended to emphasize the task of taking a careful look at what was really happening at operating nuclear plants.

* * *

A review of the documents that accumulated in the N.R.C.'s files pertaining to the nine Babcock & Wilcox reactors—they are available for inspection at the N.R.C. Public Document Room in Washington—shows what the N.R.C. would have learned if Stello's division or Denton's other specialized branches had elected to monitor these reports periodically. Although many of the documents are fragmentary, containing raw data about specific incidents and malfunctions unaccompanied by engineering analyses or commentary, they nevertheless suggest a pattern of safety problems seemingly inherent in Babcock & Wilcox designs. The problems are not, however, limited to Babcock & Wilcox reactors; the reactors of Westinghouse, Combustion Engineering, and General Electric have also experienced a wide range of safety-related difficulties, as a review of plant malfunctions suggests.

On September 16, 1978, a serious problem developed in the electrical equipment that powers and controls safety systems at Arkansas Nuclear One—Units 1 and 2. Arkansas Unit 1 uses a Babcock & Wilcox reactor, and Unit 2 uses a pressurized-water reactor designed and manufactured by Combustion Engineering. The difficulty, which could result in the disabling of all the safety equipment at both plants, was officially characterized as a case of "degradation of engineered safety features." The accident began with the sudden closing of a single valve in one of the main pipes that delivers steam to the turbine in Arkansas Unit 1, which was operating at full power. Unit 1 should have simply gone into a routine shutdown, leaving Unit 2, which was being tested before going into commercial operation, unaffected. Instead, because of a wrong interconnection of the two units, all of Unit 2's emergency safety systems were activated. Emergency diesel generators started up, valves opened or closed, pumps began pumping; the plant reacted as if there had been a major accident instead of a spurious electrical signal. Moreover, electrical-system problems at Unit 2 signaled certain valves to open prematurely—a development that caused other emergency equipment to operate out of the correct sequence. This meant that if there had been an actual crisis, such as a pipe rupture that required emergency cooling to prevent a core meltdown, the emergency cooling system at Unit 2 would not have worked. At Unit 1, meanwhile, a more routine shutdown occurred, and initially there was no evidence of deficiencies in its electrical systems. But a subsequent N.R.C. review disclosed that in a real emergency the fuses in all of Unit 1's control circuits for safety equipment could have blown, leaving the safety apparatus disabled. On February 16, 1979—five months after the accident—the N.R.C. issued a notice informing other utility companies of the "unusual sequence of events" that had occurred at the Arkansas plant. Although the N.R.C. has full legal authority to "modify, suspend, or revoke" the licenses for nuclear plants whenever it determines that a safety defect may exist, it allowed all Babcock & Wilcox and Combustion Engineering plants to continue in operation.

The commission's reference to the sequence of events at Arkansas Units 1 and 2 as "unusual" may be somewhat misleading. N.R.C. records reveal that in July 1976 a similarly widespread electrical-system failure disabled the safety equipment at the Millstone Nuclear Power Station Unit 2, in Waterford, Connecticut, on two separate occasions. That plant's pressurized-water reactor and related control equipment were designed and manufactured by Combustion Engineering. The commission had assigned one of its staff electrical experts to study this type of potential electrical problem long before either of the Millstone incidents. In his reports, the engineer concluded that circuitry failure might be a generic deficiency in the entire nuclear industry—a likely sequence of events in certain circumstances in any plant. The N.R.C. took no corrective action in response to his concern or in response to related safety problems noted soon afterward by other members of its electrical-systems branch. The reluctance of the N.R.C.'s senior management to take action on several dozen safety issues pointed out by N.R.C. engineers—a list that included the problem of mistaken operator interference with the emergency cooling apparatus—led to a decision by five licensing engineers to testify before the Senate Governmental Affairs Subcommittee on Energy, Nuclear Proliferation, and Federal Services, on December 13, 1976, at a hearing chaired by Senator John Glenn, in order to call attention to the problems, explain their seriousness, and possibly help bring about needed remedies. They told the subcommittee that the N.R.C. was "covering up" a set of safety problems that some of its own principal experts believed to be extremely grave. Their testimony stimulated none of the corrective actions they desired, although it did provoke the N.R.C. to reassign four of them two days later. The fifth engineer had already resigned from the agency in protest.

Another significant shortcoming in a nuclear plant using a Babcock & Wilcox reactor was revealed by an accident at the Rancho Seco Nuclear Generating Station, near Sacramento, on March 20, 1978. At 4:25 a.m., an operator who was replacing a burned-out light bulb on the main control panel dropped the quarter-inch bulb into the panel. The plant, which was operating at seventy percent of capacity and was supposed to be capable of coping with pipe ruptures, fires, earthquakes, and other contingencies, could not handle this seemingly trivial mishap. According to a description in a private Babcock & Wilcox report, the dropping of the light bulb caused short circuits that in turn interrupted the power supply to "a substantial portion" of the instruments in the control room that monitored key parts of the plant. The instruments included those used to control the main feedwater system, which immediately began to malfunction. Because of the instrument failure, however, the operators did not have the information they needed to determine the appropriate emergency actions. Automated equipment was available to assist in controlling the reactor, of course, but the correct automated actions were not carried out either because this equipment also failed to receive the appropriate signals from the malfunctioning instruments. To make this dangerous situation worse, the instruments started sending erroneous signals to the plant's integrated-control system—an automated master control that, among other things, regulates the main feedwater system. The integrated-control system then initiated a series of maneuvers that, according to Babcock & Wilcox, sent the reactor on a "severe thermal transient," which involved a rapid surge in pressure in the reactor—causing a relief valve to open temporarily—followed by a prolonged pressure drop. The emergency cooling apparatus was also activated. "For about the first seven minutes, the two events"—at Rancho Seco and a year later at Three Mile Island—"looked almost identical," according to Daniel Whitney, a Rancho Seco nuclear engineer. The Babcock & Wilcox report on the incident, dated August 9, 1978, written by Ivan

D. Green, the site-operations manager at the company's headquarters in Lynchburg, Virginia, concluded, "Plant operators had extreme difficulty in determining the true status ... of the plant ... and in controlling the plant because of the erroneous indications in the control room." After an hour and nine minutes, the operators finally figured out how to restore power to their instruments, and they were then able to bring the plant back under control. The Babcock & Wilcox report, intended to alert the operators of the company's other reactors to the problem, was sent out to all the utilities except Metropolitan Edison. Babcock & Wilcox officials believed that it was not necessary to inform the company of the problem, since a similar event at Unit 2 on March 23, 1978, had already been discussed with Met Ed officials.

* * *

The copy of the Babcock & Wilcox report on the Rancho Seco accident sent to the Davis-Besse Nuclear Power Station in Ohio was ultimately seen by James S. Creswell, a thirty-five-year-old reactor inspector in the Midwest regional office of the N.R.C.—in Glen Ellyn, Illinois, near Chicago—whose duties included overseeing the startup of Davis-Besse. Although reports of the Rancho Seco accident, and of accidents that had occurred at the Davis-Besse plant, had been sent to the N.R.C., these reports, far from stimulating an N.R.C. review of Babcock & Wilcox plants, had received scant official attention. Creswell was deeply concerned about the accidents, however, and about several other problems that had come to his attention at Davis-Besse. On his own initiative, he raised a number of questions about Babcock & Wilcox reactor problems—not only with plant officials but also with officials of the N.R.C. He felt that there might be safety flaws inherent in the Babcock & Wilcox equipment—a view not shared by senior N.R.C. officials.

One of the accidents that worried Creswell had occurred at Davis-Besse on September 24, 1977—an accident that, it is clear in retrospect, bore a striking resemblance to the one at Three Mile Island eighteen months later. At Davis-Besse, the main feedwater system failed—as it did at Three Mile Island—and then a pressure-relief valve opened and, because of a missing electrical relay, cycled open and closed about nine times, finally jamming in an open position. (It is still not known whether the pressure-relief valve at Three Mile Island stuck open because of an electrical-control problem or for some other reason, since the continued high radiation levels inside the containment building have made a physical inspection impossible.) The operators at Davis-Besse, like the operators at Three Mile Island, did not immediately realize that the relief valve was open. At Davis-Besse, it took twenty-two minutes to discover the open relief valve; at Three Mile Island, it took about two hours and twenty minutes. At both plants, the pressure in the reactor's cooling system increased and then decreased uncontrollably. Part of the emergency feedwater system malfunctioned at Davis-Besse, as it did at Three Mile Island. The Three Mile Island plant, however, experienced a complete, rather than a partial, initial failure of its emergency feedwater equipment. Finally, at both plants, the operators mistakenly shut off the emergency cooling pumps—as a result of their erroneous interpretation of the pressurizer-water-level-instrument readings. The main difference between the two accidents—and it is a major reason that the Davis-Besse accident was much less severe and did not result in core damage, as the accident at Three Mile Island did—was that the Davis-Besse reactor was operating at only nine percent of its full power, whereas the reactor at Three Mile Island was operating at ninety-seven percent.

Worrying about the adequacy of safety systems was supposed to be done by the hundreds of safety analysts at the N.R.C.'s headquarters, not by a field inspector like Creswell, who was expected to be more of a bookkeeper than an engineer. His principal assignment was to monitor the paperwork connected with a utility company's quality-assurance programs and to review other aspects of its administrative procedures. The N.R.C. had elected not to have its inspectors carry out comprehensive physical inspections of the actual equipment and components installed in nuclear-power plants—not to test equipment performance, or periodically check sensitive pipes for cracks, or determine whether specific components had been built and installed in accordance with N.R.C. safety requirements. But Creswell, in the light of the safety problems that he saw at Davis-Besse, was not content to stick to his narrow formal assignment and overlook what he took to be the possible generic safety ramifications of the September 24, 1977, accident and the problems related to it. Nor was he satisfied that he could rely on the N.R.C.'s determination that the plant's Babcock & Wilcox equipment was adequately designed or on the utility company's "self-evaluation," which stated that the various "incidents" at the plant were no cause for concern.

"The problems at Davis-Besse just bugged me," Creswell said in an interview a few months after the Three Mile Island accident. His pursuit of these matters, he recalled, began in earnest after a routine three-day inspection that he made at the plant starting on December 6, 1977. The focal point of this particular inspection was an occurrence at the plant a week before, on November 29. A temporary power failure had caused the plant to shut down suddenly—a relatively common experience for any type of electric-generating station. As Creswell reviewed the records of the incident, however, he became concerned about the strange behavior of the pressurizer-water-level instrument during the incident. This is a device that reactor operators use to tell how much water is in the reactor. During the November 29 incident, the pressurizer-water-level instrument indicated that the pressurizer had no water in it. With that instrument showing "zero," the operators had no indication—not even this indirect one—of how much water was in the reactor. Although the incident ended quickly, with the reactor satisfactorily stabilized, Creswell began to worry how operators of pressurized-water reactors could possibly cope with plant emergencies if they didn't know how much cooling water was in the reactor.

He talked to Davis-Besse officials about the pressurizer-instrument problem during his inspection of the plant in December 1977 and several times afterward, during ten inspections he made in 1978. As he looked over the records for information about how the pressurizer instrument on Babcock & Wilcox reactors behaved, he gathered more details about the September 24, 1977, incident—the one that so strongly prefigured the accident at Three Mile Island. He learned that the operators at Davis-Besse, on the basis of the pressurizer-water-level-instrument readings, had prematurely shut off the emergency-cooling-system pumps—even though the reactor's cooling water was still being lost through the stuck-open relief valve. (This key fact had been omitted in the "licensee event report" on the accident that the company submitted to the N.R.C. on October 7, 1977.) Creswell formally asked Davis-Besse officials to review the operating procedure according to which the control-room personnel had deliberately disabled the emergency cooling system, telling them that the action was contrary to the way in which the system was intended to work. Company officials were not responsive, Creswell says; they replied that a review of the matter was "in process" and, moreover, that it was really outside the scope of his responsibilities as an inspector, being a subject more properly addressed by safety analysts at N.R.C. headquarters.

N.R.C. headquarters, it turned out, had inquired into the September 24 accident. Denwood Ross, a senior official in the Division of Systems Safety, dispatched Gerald Mazetis, a reactor-systems engineer, to Davis-Besse to attend a special meeting on September 30, at which the accident was to be reviewed. In his report on the meeting, Mazetis noted that the operators had disabled the emergency cooling equipment and that there were "many questions . . . to be addressed." The accident was then reviewed on October 3, at a meeting in the office of Dr. Roger Mattson, director of the Division of Systems Safety, which was also attended by Karl Seyfrit, an assistant director of the Office of Inspection and Enforcement. This office assumed responsibility for investigating the event further—a responsibility confirmed in a follow-up memorandum from Ross. Item 2 in the memorandum, which was dated October 20, 1977, states, "The operator's role in participating in the event should be related. . . . The operator's decision to secure [emergency-cooling-pump] flow based on pressurizer level indication should be explained." There is, however, no evidence that a satisfactory follow-up investigation was ever carried out. Creswell himself, during his continuing review of the September 24 incident, called N.R.C. headquarters asking for documentation on how the matter was resolved. He was told, he says, that there was none.

Instead of encouraging Creswell to investigate the safety issues raised by the accident at Davis-Besse, his superiors at the N.R.C. were, he says, "quite negative" about his continuing work on these matters. It has been tacit N.R.C. policy, Creswell explains, not to aggressively pursue safety questions that could have a major adverse economic impact on the nuclear-power industry. There is, of course, an obvious trade-off between the safety and the economics of nuclear power—one that the N.R.C. has had to finesse constantly. N.R.C. officials, in view of the long-standing federal policies favoring nuclear-power expansion as well as their own personal and professional commitment to the program, have often been sympathetic to the industry in deciding where to draw the line. This sympathy has been especially in evidence when a safety concern has arisen that might necessitate the shut-down of an operating nuclear plant—something the N.R.C. is extremely reluctant to order. The agency's tolerant attitude toward newly discovered safety problems has also been influenced by a general complacency, after years of seeing such problems come and go, about the risk that one of them might grow into a serious accident. The N.R.C. had developed what Creswell calls the "mind-set" that such accidents just "couldn't happen." Creswell, who in 1978 had been with the agency for only two years, had yet to absorb the official outlook, and could not so easily dismiss the safety problems he saw at Davis-Besse.

For fourteen months, despite the opposition of his superiors, Creswell did everything he could think of within the N.R.C. to call attention to the problems of Babcock & Wilcox reactors. He pestered Davis-Besse officials for more information about the accidents there until they complained to his superiors; he asked for an investigation, under the commission's regulations covering the disclosure of safety defects by the industry, to determine whether Babcock & Wilcox might have withheld information about the safety of its plants; he wrote memos that urged further investigations of various technical issues; and, generally, he pulled every bureaucratic lever available to him. He got nowhere. Finally, he decided to press his concerns far outside the normal channels. In a memorandum dated January 8, 1979, he set out his views on the overall problems at Babcock & Wilcox plants, and formally asked to have this memorandum submitted to the N.R.C. licensing boards considering authorizations for additional Babcock & Wilcox plants. This maneuver would, in effect, make public the concerns he had expressed within the N.R.C. because all correspondence received by the licensing boards becomes public. Creswell hoped that his memo might attract the attention of technically competent critics outside the N.R.C., who would put pressure on the commission to resolve the problems. Even though official procedures specified that his superiors must transmit his memo to the licensing boards, they were hesitant to do so, since this would mean that the official N.R.C. safety evaluation of Babcock & Wilcox plants would be called into question in public licensing hearings. Creswell's memos were distributed to senior officials throughout the N.R.C., but it was not until March 29, 1979—the day after the Three Mile Island accident began—that they were sent to the licensing boards.

The manner in which the N.R.C. disposed of Creswell's concerns can be deduced from the "Weekly Information Report—Week Ending December 29, 1978," which was sent directly to the five commissioners by the N.R.C. staff, and which refers to one of the major problems Creswell raised. At issue was the performance of the emergency feedwater system in the Davis-Besse plant—the same piece of safety equipment that failed at Three Mile Island three months later and that had partly failed at Davis-Besse on September 24, 1977. If the main feedwater system fails and the emergency feedwater systems also fail, no satisfactory method for cooling the reactor is available. The uranium fuel could then overheat dangerously, as happened during the accident at Three Mile Island Unit 2. (On some pressurized-water reactors of German design, extra "residual-heat-removal systems" have been installed to cool the reactor in the event of total failure of the main feedwater and emergency feedwater systems. American reactor designers, with the acquiescence of the N.R.C., have chosen not to install this expensive additional equipment; if they had, the accident at Unit 2 would have easily been controlled.)

There was another aspect to the feedwater-control problem noted by Creswell—something that has long been known to reactor-safety experts: too much emergency feedwater flow can be as destabilizing as too little. If more emergency feedwater than is required to remove the heat produced by radioactive decay is pumped into the steam generators, the high pressure in the primary cooling system can be suddenly reduced. Some of the primary cooling water could then start to boil and flash into steam, posing two problems for the reactor's hot uranium core. First, if the core is filled with steam or a mixture of steam bubbles and water instead of just water, the uranium fuel is not as adequately cooled, because water removes heat from the surface of the fuel rods better than steam does. Second, if a large steam bubble develops inside the reactor, it could block part of the piping system and interfere with the circulation of water through the reactor; this could make it difficult to keep the core temperature under control. What is needed, in short, is just the right balance of emergency feedwater flow to keep the pressure and the temperature of the primary reactor system under control. The Rancho Seco electrical failure led to incorrect emergency feedwater flows, and to a dangerous pressure decrease in the reactor system. Similar accidents had occurred at Davis-Besse. Creswell wondered whether there were generic defects in Babcock & Wilcox reactors that allowed these accidents to happen. Babcock & Wilcox denied that any such problems existed.

On December 22, 1978, in response to Creswell's persistent criticism, the owners of the Davis-Besse plant had submitted to the N.R.C. an "Additional Safety Evaluation" prepared by Babcock & Wilcox. This report argued that the plant's emergency feedwater systems were fully capable of dealing with any contingency, including the possibility that "overcooling transients" from "excessive" feedwater might jeopardize plant safety. The report asserted that Creswell had raised "no unreviewed safety question" and no other matter that had not already been satisfactorily taken into account in the design of Davis-Besse and all the other Babcock & Wilcox plants. Point-by-point responses by the company to Creswell's concerns included the declaration that steam bubbles—or "voids," as the company called them—"will not collect in reactor coolant piping and no flow blockage will occur." Further analyzing the contingencies, the company said that "during the course of" such incidents "no core problems develop." The company engineers added, "We conclude that no safety problem exists.... No adverse consequences ... have been shown and, therefore ... no concerns to the safe operation of the plant."

A day after receiving the Babcock & Wilcox report, the N.R.C.'s Division of Operating Reactors concluded that no action was required. The swift determination in favor of the owners of the Davis-Besse plant was, for this particular N.R.C. division, a routine move—one that it could make without detailed deliberations, and one that squared with the outlook of Victor Stello, its director at the time. Stello, a large man with a broad, swarthy face and a booming voice, is one of the most influential members of the N.R.C.'s top management. He is also a controversial figure, and some of his colleagues have privately criticized his optimistic views on the adequacy of nuclear-plant safety precautions. Like a number of other key N.R.C. officials, Stello was a member of the upper echelon responsible for many of the basic safety policies under which Babcock & Wilcox reactors and dozens of others received their official safety certifications. For these officials, to acknowledge a serious safety problem with one of these units is to raise questions about the correctness of their own past decisions. To admit even broader safety problems, such as generic defects in an entire set of previously licensed plants, is to concede major blunders. Such concessions—what bureaucrats delicately term "backpedaling"—are all the more difficult in the light of the growing nuclear controversy and what one N.R.C. technical-review group in 1978 termed the consequent "siege mentality" of some senior members of the N.R.C. staff.

"After the regional inspector questioned the licensee's determination that no unreviewed safety question was involved," the "Weekly Information Report—Week Ending December 29, 1978" said, the Division of Operating Reactors "requested a copy of the licensee's evaluation and, subsequently, additional information." The report continued, "The additional information was submitted on December 22, 1978. After review on December 23, 1978, D.O.R. concluded that because no fuel damage would occur . . . no unreviewed safety question was involved and no licensing action was required." This summary dismissal of Creswell's concerns left Davis-Besse, Oconee Units 1, 2, and 3, Rancho Seco, Crystal River Unit 3, Arkansas Nuclear One—Unit 1, and Three Mile Island Units 1 and 2 licensed for full-power operation with the approval of Stello's division and the concurrence of the entire senior management of the commission.

Still, Creswell, who was now confronting the heads of staff and no longer merely his immediate supervisors, continued to challenge the N.R.C.'s conclusions about the safety of Babcock & Wilcox plants. It was at this point that Creswell had decided to write the memorandum that would go to the N.R.C. licensing boards and thus, he hoped, put some public pressure on the agency. His January 8, 1979, memo pointed out "certain issues [that] have come to my attention" during "the course of my inspections at Davis-Besse." Creswell mentioned six problems, including the feedwater-control problem and a related one—having to do with the pressurizer-water-level instrument. His memo went from office to office inside the N.R.C. By March 6, 1979, it had been delivered not only to Stello but also to Harold Denton, the director of the Office of Nuclear Reactor Regulation; Edson Case, Denton's deputy; Darrell Eisenhut, deputy director of the Division of Operating Reactors; Brian Grimes, assistant director for engineering and projects in the Division of Operating Reactors; Steven Varga, chief of Licensing Branch No. 4; Dudley Thompson, executive officer for operations support in the Office of Inspection and Enforcement; and other senior N.R.C. officials.

"I don't think I had read Creswell's memo at that time," Denton, the most senior N.R.C. official who received it, said in an interview not long after the Three Mile Island accident. Denton conceded that he and other senior N.R.C. officials had "sort of made a misallocation of resources" by not studying the memo. "Creswell was right on in many of his ideas, but for some reason, the system didn't elevate Creswell's concern to a priority where it was recognized above many other ones that were thought to be high priority," he said. Denton strongly resisted any suggestion that the agency was negligent in its handling of Creswell's memos, saying that "whatever the accident" that might strike a nuclear plant, there would always be a "paper trail" in the agency's files showing that somebody had previously worried about that kind of thing. Denton offered no specific reasons why Creswell's memos and such seemingly serious events as the Davis-Besse and Rancho Seco accidents failed to attract his attention or that of others in the senior management of the N.R.C. Nor did he suggest why the officials who received Creswell's memos took no action to see that the safety problems pinpointed in them were corrected. It was only a few weeks after they received Creswell's January memo that all these key officials hurriedly assembled in the N.R.C.'s emergency-response center, in Bethesda, Maryland, or went off in helicopters to Harrisburg, Pennsylvania, to see Creswell's diagnosis of the Babcock & Wilcox safety problems proved right. "Well, in retrospect, Creswell was pretty much on target," Denton said. "It just wasn't recognized."

Creswell was discouraged but not defeated by the response his reports received from the N.R.C.'s top officials. Confident of his technical ground—and relying on a background in nuclear engineering gained during five years in the operations division of the Tennessee Valley Authority in Chattanooga, Tennessee—Creswell decided to go directly to the N.R.C. commissioners. In a maneuver that a conventional bureaucrat would regard as paralleling an attempt by a corporal to make an appointment with the secretary of defense, Creswell set up a meeting for himself with Commissioners John Ahearne, a physicist and former Defense Department official, and Peter Bradford, a lawyer and former chairman of the Maine Public Utilities Commission. He thought that, of the five commissioners, these two, who were the most recently appointed and had had no association with the A.E.C., would give him the fairest hearing. He had to take a day off from work to go to Bethesda for the meeting, and he had to make the round-trip flight from O'Hare International Airport in Chicago at his own expense. The meeting went well, he thought: the two commissioners agreed to take a close look at the issues he had raised. The following week, Ahearne rewarded Creswell's fourteen months of persistence: he prepared a memo to the N.R.C. staff asking for a report on some of the safety questions affecting Davis-Besse that Creswell wanted answered. Ahearne's memo—the only positive response by any influential member of the N.R.C. to the concerns expressed by Creswell—was too late: the accident at Three Mile Island Unit 2 had occurred the day before.

Six months later, Creswell was given a special N.R.C. award of four thousand dollars in recognition of his diligence in trying to alert the agency to important safety problems. Stello and Denton, whose two divisions ignored Creswell's warnings, also received awards—of ten thousand and twenty thousand dollars, respectively—for their part in the N.R.C.'s efforts to respond to the accident. The appropriateness of these latter awards has been publicly questioned by Mitchell Rogovin of the Special Inquiry Group. Rogovin has said that it would have been more fitting for "the senior officials of the N.R.C. to owe the government money," given their performance. They have acknowledged no such debt and remain the most important heads of staff inside the N.R.C. Creswell, on the other hand, recently submitted his resignation, explaining in an interview that he thought there were "greater opportunities" for him outside the agency.

Although Creswell did not know it at the time, he was not alone in his efforts to alert the N.R.C. to the problems that might undermine the safety of Babcock & Wilcox reactors. In late 1977, at practically the same time that Creswell had begun raising the issue, Carlyle Michelson, a senior nuclear engineer at the Tennessee Valley Authority who worked on safety analysis, was completing a detailed technical analysis that emphasized some of the same basic concerns. Michelson had been working on a review of the T.V.A.'s Bellefonte Nuclear Plant, which is now under construction in Scottsboro, Alabama, and which incorporates a Babcock & Wilcox reactor. Along with Jesse Ebersole, a retired T.V.A. engineer, who began serving as a member of the N.R.C.'s Advisory Committee on Reactor Safeguards in 1976, Michelson had been questioning a number of nuclear-safety assumptions—an activity not entirely welcomed by the T.V.A., which is heavily committed to nuclear power—and had been particularly concerned about small loss-of-coolant accidents: leaks in the reactor's primary cooling system caused by ruptures of small pipes or by a relief valve stuck open. Michelson and Ebersole felt that such accidents had been overlooked, or only superficially studied, because of all the emphasis on hypothetical loss-of-coolant accidents caused by ruptures of large pipes. The prevailing assumption behind the official N.R.C. safety analyses was that this emphasis was "conservative," since if a plant could cope with a big loss-of-coolant accident, it would presumably have no difficulty in coping with a small one.

In January 1978, Michelson finished a thirty-four-page technical report on how small loss-of-coolant accidents might affect Babcock & Wilcox reactors—a document that demonstrates just how refined and specific the warnings had been about the type of accident that struck Three Mile Island on March 28, 1979. The report specifically addressed the "heat-removal problem" that could arise during small loss-of-coolant accidents in large Babcock & Wilcox reactors, and its conclusions were quite straightforward: it might be extremely difficult to keep the reactor adequately cooled during such emergencies. There were "one or more impediments to decay heat removal" in Babcock & Wilcox plants that would prevent the emergency cooling equipment from performing as intended. Among these "impediments" was the formation of a steam bubble that could lodge inside various parts of the reactor's cooling system and prevent the circulation of cooling water. Babcock & Wilcox had insisted, in its response to Creswell's criticism, that such a steam bubble could not form, but it did form during the Three Mile Island accident. Another serious problem was that the pressurizer-water-level instrument was "not a correct indicator of water level over the reactor core" and that there might not be adequate "emergency operating procedures and operator training" at Babcock & Wilcox plants to allow the operators to handle serious reactor-cooling problems. The operators, Michelson predicted, might mistakenly turn off emergency cooling equipment because of misleading pressurizer-instrument readings—exactly what happened on March 28, 1979. Michelson recommended that studies be carried out to get more information on the "stability" of Babcock & Wilcox cooling equipment and on the "adequacy of instrumentation and components" used in these systems. In a letter dated April 7, 1979—ten days after the Three Mile Island accident—the Advisory Committee on Reactor Safeguards asked the N.R.C. to carry out some of Michelson's recommendations. Some of the "impediments" to heat removal in Babcock & Wilcox reactors that had been described in Michelson's report had by then been demonstrated.

* * *

In September 1977, Jesse Ebersole gave a handwritten draft copy of Michelson's report to Sanford Israel, a mechanical engineer in the Reactor Systems Branch of the N.R.C. Ebersole says that he wanted to find someone in the N.R.C. who would be "receptive" to the report and "competent to analyze it." In November of that year, Ebersole again raised some of the issues identified by Michelson at an Advisory Committee on Reactor Safeguards meeting, and his questions were forwarded to Israel's branch at the N.R.C. A few months later, Israel wrote a two-page memo that was sent to other members of the Reactor Systems Branch. The memo, dated January 10, 1978, was distributed over the signature of Thomas Novak, chief of the Reactor Systems Branch—a standard N.R.C. procedure under which the branch chief adds an endorsement to the work of a staff engineer. Israel's memo noted that there were some peculiarities in the pressurizer design at Babcock & Wilcox plants—specifically, in the way this tank was connected to the piping coming out of the reactor. Israel stated that "under ordinary circumstances" these design features were "inconsequential" but that "under upset conditions," such as a "prolonged relief valve opening," the loss of cooling water from the reactor "may not be indicated by pressurizer level," and noted, "This situation has already occurred at Davis-Besse 1 when a relief valve stuck open." A "full" reading on this instrument, he noted, interpreted by the control-room operators in the standard way, could lead them to "terminate" emergency-cooling-system pumps when it was imperative to keep them running. The memo suggested that it was a matter of some urgency to review the pressurizer-design requirements and operating procedures for new Babcock & Wilcox plants to correct this problem. However, the memo made no mention of corrective action for the nine Babcock & Wilcox plants already in operation. This was not a casual oversight on Israel's part. The nuclear industry vehemently opposes new safety requirements—especially retroactive ones entailing basic changes in plants that have already been licensed and put into commercial service. Since it is much easier to make changes when a proposed plant is still on the drawing boards, the N.R.C. tries to limit mandated safety improvements to new plants, thereby sparing the industry what might otherwise become a never-ending program of costly postconstruction modifications. By sidestepping the question of the operating Babcock & Wilcox reactors, Israel was simply following standard agency practice. He explained in an interview that he had written the memo simply to remind N.R.C. engineers to look at the pressurizer-instrument problem the next time a new Babcock & Wilcox plant came up for safety review. Israel says that he got "no feedback" from the N.R.C. officials who received it. Certainly none of them took any action—not even the simple precaution of informing the operators of the nine Babcock & Wilcox reactors that a deficiency had been found in one of the key control-room instruments used during emergencies.

The most senior N.R.C. official to receive a copy of Israel's memo was Denwood Ross, who was then assistant director of the Division of Systems Safety. Among other duties, Ross supervised an engineering staff responsible for studying how well the emergency cooling systems in each nuclear plant would function during major emergencies. He has a doctorate in nuclear engineering from Catholic University in Washington, and, like most other top members of the N.R.C. staff, worked for the A.E.C., which he joined in 1967. Ross took no action in response to Israel's memo—a circumstance that is particularly puzzling in the light of Ross's earlier apparent concern over the Davis-Besse accident of September 24, 1977. It was Ross, after all, who had promptly sent Gerald

Mazetis to Davis-Besse to investigate that accident and had written the memo of October 20, 1977, asking the N.R.C. Office of Inspection and Enforcement to find out why the operators at Davis-Besse had shut off the emergency cooling pumps. Israel's memo, on its face, provided Ross with a precise technical answer to his question: the pumps were shut off because of the misleading pressurizer-water-level-instrument readings. Nevertheless, Ross did nothing when he received Israel's memo—an "oversight" that he cannot readily explain, he said in an interview. "In retrospect, I'm not sure why the memo wasn't sent to the Division of Operating Reactors," he said, indicating what he thought was "the right thing to have done with it," since Israel's finding affected the safety of the Babcock & Wilcox reactors then in operation. Ross suggested that his failure to forward the memo through the appropriate channels was—in some measure, at least—a bureaucratic slipup. "The Davis-Besse analysis was under the jurisdiction of the Office of Inspection and Enforcement, and I guessed that if anything needed to be done they would do it," he said. Ross candidly admitted, however, that his own overall attitude toward nuclear-plant safety—he had been one of the staunch defenders of the adequacy of the safety rules developed by the A.E.C. and used by the N.R.C.—was also involved in his failure to pursue Israel's memo further and to get to the bottom of the Davis-Besse accident.

> We screwed up—and I mean by "we" the nuclear community, the vender [Babcock & Wilcox], and the N.R.C. Davis-Besse was ample warning, and if we had paid ample attention to it Three Mile Island could have been prevented. If you want to use the term "complacency" to describe our behavior, I won't quibble with you, but the term I'd rather use is "mind-set," or "attitude." I'd had the attitude that reactors were fairly forgiving, in the sense that they could withstand a lot of problems without having those problems turn into serious accidents. I don't feel that way anymore.

* * *

Babcock & Wilcox was immediately notified of the incident at Davis-Besse on September 24, 1977—the event that so strikingly foreshadowed the accident at Three Mile Island. Obviously, the company's large scientific and engineering staff was intimately familiar with the features of its nuclear-reactor systems, and therefore was much better equipped than outsiders—such as Creswell, Michelson, and Israel—to assess the implications of the Davis-Besse accident. Babcock & Wilcox records subpoenaed during the official investigations of the Three Mile Island accident show that the company's safety experts quickly appreciated the significance of the Davis-Besse episode and readily understood that remedial action was necessary to prevent an accident like the one at Three Mile Island. Babcock & Wilcox, however, according to their records, never told the operators of the nine licensed Babcock & Wilcox reactors about its findings.

On September 25, 1977—the day after the accident—Joseph Kelly, an engineer in the plant-integration unit, which coordinates the design of the various components used in Babcock & Wilcox reactor systems, was sent to Davis-Besse from the company's headquarters in Lynchburg, Virginia. No office or individual within the company had formal responsibility for analyzing the accidents that occurred at plants with Babcock & Wilcox reactors, according to the investigation by the President's Commission on the Accident at Three Mile Island, and the company itself did not usually even receive the full reports on such accidents that its customers filed with the N.R.C. Babcock & Wilcox is preoccupied largely with the business of designing and selling new nuclear reactors, and it devotes little time or money to looking over its shoulder, so to speak, at the equipment it has already sold. Still, the problem at Davis-Besse was such a complicated and unusual one that the company's customer-service department, which gives operating instructions to the owners of Babcock & Wilcox reactors, asked to have someone sent to investigate it. Kelly returned to the Lynchburg headquarters after about two days at the Davis-Besse site, and briefed a group of some thirty Babcock & Wilcox employees on the basic details of the accident. In his audience were the heads of several major company departments and the vice president of the Babcock & Wilcox Nuclear Power Generation Division, John MacMillan. Kelly's report mentioned the problems with the stuck-open relief valve, the misleading indication on the pressurizer-water-level instrument, and the fact that the operators had shut off, or throttled, the emergency-cooling-system pumps. It did not occur to Kelly until after the meeting, though, that this operator interruption of the emergency cooling pumps had serious safety implications—a fact that was pointed out to him by Bert Dunn, the manager of emergency-cooling-system analysis. Dunn explained to Kelly that there were "scenarios"—such as having a reactor at full power rather than at the nine percent power level of the Davis-Besse plant—in which mistaken interference with the emergency cooling equipment could lead to serious difficulties. This was the first of several discussions between Dunn and Kelly on these potential difficulties.

Senior Babcock & Wilcox officials took no formal action after Kelly's briefing, although for the next several months the matter was the subject of memorandums, meetings, and conversations involving officials from at least seven different departments of the company. Kelly himself went to the training-services section in October to discuss the instructions that were being given to plant operators about the proper operation of the emergency cooling pumps. John Lind, the chief instructor, assured Kelly that the correct instructions were being given, but he could not explain why the operators at Davis-Besse had shut the pumps off on September 24. Kelly also continued his discussions with Dunn and prepared a formal memorandum, dated November 1, 1977, in which he set out some of his concerns. This memo was sent to seven Babcock & Wilcox officials. Kelly briefly summarized the Davis-Besse accident of September 24, and noted that a similar event had occurred at the same plant on October 23, in which the operators deliberately prevented the emergency cooling pumps from operating. "Since there are accidents which require the continuous operation" of these pumps, Kelly wrote, "I wonder what guidance, if any, we should be giving to our customers on when they can safely shut the system down following an accident? . . . I would appreciate your thoughts on this subject." He himself proposed that a two-sentence guideline be issued to reactor operators admonishing them, for example, not to prevent the activation of the emergency cooling pumps "under *any* conditions except a normal, controlled plant shutdown." No action was taken on Kelly's memorandum, and none of the seven officials to whom it was sent responded.

Three months later, after discussing the matter further, Kelly and Dunn decided to try to compel some action. Dunn, the senior of the two, wrote a memo, dated February 9, 1978, to James Taylor, the manager of the company's licensing section, who was supposed to inform the N.R.C. whenever any problem was discovered that could create a substantial safety problem at an operating nuclear-power plant. (Full compliance by the nuclear industry with this disclosure requirement is an essential element in the N.R.C.'s scheme of industry self-regulation.) Dunn sent copies of his memo to ten other company officials, two of whom had received Kelly's memo on the subject. Dunn, a physicist who had worked on emergency-cooling-system analysis for the company for almost a decade, noted at the outset of his memo that he had "a serious concern" about operator interference with the proper functioning of the emergency cooling system. During the September 24 accident at Davis-Besse, Dunn wrote, the operators "terminated" the emergency cooling pumps because of "an apparent system recovery indicated by high [water] level within the pressurizer," even though, according to Dunn's subsequent analysis, the reactor was still losing its cooling water. "I believe it fortunate that [Davis-Besse] was at an extremely low power" when the event occurred, he continued:

> Had this event occurred in a reactor at full power . . . it is quite possible, perhaps probable, that core uncovery and possible fuel damage would have resulted. The incident points out that we have not supplied sufficient information to reactor operators. . . . I believe this is a very serious matter and deserves our prompt attention and correction.

He did not get a response from Taylor, who told Dunn later—after the Three Mile Island accident—that his memo must have been "misdirected," because it had not been submitted on the appropriate form to qualify as a "safety concern."

A copy of Dunn's wayward memo had been sent to the customer-service department, which held a discussion in mid-February of 1978 in which Dunn agreed to a proposed set of new guidelines to be sent to plant operators, telling them how to operate the emergency cooling pumps. Dunn believed that the guidelines took care of his concerns, and so informed Taylor in another memo, dated February 16, 1978. Dunn then considered the matter closed. Unfortunately, the customer-service department had some second thoughts about the proposed guidelines, and they were never sent to the operators of Babcock & Wilcox plants—a fact that Dunn has said he did not learn until after the Three Mile Island accident.

The fitful response by Babcock & Wilcox officials to Dunn's memorandums continued throughout 1978 and into early 1979—long after Dunn had turned his attention to other matters. During the spring and early summer of 1978, for example, the customer-service department sought the assistance of analysts in the plant-integration unit, asking, in particular, for a revision in the operating procedures that were leading plant operators to shut off the emergency cooling pumps. By August 1978, there was still no response from the plant-integration group, and Donald Hallman, of the customer-service department, sent yet another memo trying to direct some attention to the problem. A memo on August 3, 1978, noted that "to date" the customer-service department "has not notified our operating plants" to change their emergency-cooling-system procedures, as Dunn's memos had suggested. Bruce Karrasch, the manager of plant integration, skimmed the memo and, considering it a routine matter, forwarded it to one of his deputies, who does not remember receiving it. Hallman says that he attempted to discuss the matter with Karrasch several times in the period between August 3, 1978, and March 15, 1979, but had no success. Hallman has said that he did not himself assign a high priority to the issue, and Karrasch has told federal investigators, "All I can remember is that in the fall of 1978 things were very, very busy . . . and my attention and the whole group's attention [was given] to what were perceived to be higher-priority matters." The entire problem, company officials have told investigators, simply fell between the cracks.

Babcock & Wilcox internal documents, which show that company officials knew about the unresolved problem of operator interference with emergency cooling equipment, raise the question whether the company complied with the N.R.C. requirement covering the disclosure of safety-related data. The company, after all, told neither plant operators nor the N.R.C. of the problem, though N.R.C. regulations state that all companies doing business in the nuclear power industry are required "to immediately notify" the N.R.C. whenever they obtain "information reasonably indicating" that any licensed nuclear plant "fails to comply" with N.R.C. safety regulations or "contains a defect" that could create "substantial safety hazards." (The disclosure requirement is set forth in Part 21 of Title 10 of the Code of Federal Regulations, in a section entitled "Reporting of Defects and Noncompliance.")

In light of the Babcock & Wilcox internal memos, the N.R.C.'s Office of Inspection and Enforcement conducted an investigation and determined that the company had violated the disclosure provision. A fine of a hundred thousand dollars was imposed on Babcock & Wilcox. The company denied the charges but said it would pay the fine in order to avoid the time and expense of fighting the N.R.C. The company did announce, however, that it was instituting new procedures for the "processing of safety concerns" so that in the future it would comply with the N.R.C. disclosure requirement. According to the new procedures, safety concerns were henceforth to be reported by employees on company form BWNP-20208. The concerns would then be logged and numbered and would be distributed within the company, as specified in a twelve-step "flow chart," in accordance with Management Directive 205 T4.4, dated April 10, 1980. At the end of all this, the N.R.C. would be notified, if the company found it appropriate to do so.

In addition to informing plant operators that the pressurizer-water-level readings might be a misleading indicator of the amount of water inside the reactor, there were other, more positive steps that Babcock & Wilcox could have taken to help the plant operators handle accidents with the reactor-cooling system. One of the most obvious steps, of course, would have been to install instruments that directly measured how much water was inside the reactor. The operators would then know, without ambiguity, whether the essential safety goal of keeping the hot uranium fuel adequately covered with water was being achieved. Reliable equipment to measure the water level in the reactor, however, had never been installed on any of the pressurized-water reactors in the United States—including those manufactured by Westinghouse and Combustion Engineering as well as by Babcock & Wilcox. "This problem of what we call vessel-level indication for pressurized-water reactors has been kicked around probably at least for the last ten years," Carlyle Michelson has said. During normal operation, when the reactor is always full of water, he admitted, such a device has no real point, but during a loss-of-coolant accident "it would be awfully nice to know" the water level in the reactor—yet, he added, "it was never deemed to be quite nice enough to know to justify the cost."

* * *

A look at the operating records of the nine nuclear power stations with Babcock & Wilcox reactors underscores the urgency with which the warnings of Creswell, Michelson, Israel, Kelly, and Dunn should have been treated. Summary statistics now available from the N.R.C. show that a virtual epidemic of serious cooling-system difficulties had been occurring at these nine reactors since the first of them—at the Oconee plant in South Carolina—went into operation in 1973. Such data, however, were not systematically compiled and analyzed by the N.R.C. or by Babcock & Wilcox, and no official notice was taken of the disturbingly high frequency of these events. One particular statistic, which would have helped the N.R.C. evaluate the problem of small loss-of-coolant accidents arising from relief-valve openings, was the actual rate at which the relief valve was being forced to open at the nine plants. This valve is supposed to open only infrequently—when it becomes necessary to reduce excessively high pressure inside the reactor. The reason for this is obvious: the valve's function is to remove vital coolant from the reactor, and this is hardly something that a prudent designer would wish to have happen very often. Yet, according to belatedly assembled N.R.C. statistics, the relief valves on the nine Babcock & Wilcox reactors had opened on at least a hundred and fifty occasions before the Three Mile Island accident. (The official bookkeeping is not precise, since the utility companies operating the plants did not always notify the N.R.C. when relief-valve openings occurred.) Each of the plants experienced, on the average, five relief-valve openings per year of operation.

The high rate of relief-valve openings is attributable to design weaknesses inherent in Babcock & Wilcox reactors. Unlike the pressurized-water reactors designed by Westinghouse and Combustion Engineering, Babcock & Wilcox equipment has a pronounced susceptibility to sudden increases in the pressure of its primary cooling water whenever a plant has to be shut down quickly. This occurs for two principal reasons. First, the steam generators in Babcock & Wilcox reactor systems contain less water than those of the two other companies. This means less cooling capacity and a greater likelihood, during a sudden plant shutdown, of an increase in the temperature and the pressure of the primary cooling water—which, of course, is the condition that causes the relief valve to open. Second, there was a time delay inherent in the control system that shut off the Babcock & Wilcox reactors whenever common electric-generating-plant malfunctions—such as problems with the turbine—necessitated an unscheduled reactor shutdown. Because a Babcock & Wilcox reactor would continue to operate for several seconds after the turbine had failed, the excess energy in the reactor, which was no longer being used by the turbine, caused the pressure of the primary cooling water to increase. The excess pressure, in turn, caused the relief valve to open. (The reason for the time delay was economic: Babcock & Wilcox designers wanted to avoid a costly reactor shutdown every time a minor malfunction occurred in some other part of the plant.)

Not only had the relief valves at Babcock & Wilcox plants opened with uncommon and undesirable frequency but they had also—on at least nine occasions before the Three Mile Island accident—opened and stuck open. (The valves themselves—which weigh about a hundred and seventy-five pounds each and cost some thirty thousand dollars—are made by specialty valve manufacturers, such as Industrial Valve Operations of Dresser Industries, not by Babcock & Wilcox.) These events were all the more serious because in some cases plant operators remained unaware for appreciable periods that, as a consequence, the reactor was losing its coolant. As the operating records indicate, various conditions were responsible for the failure of the valves to reclose properly. In the Davis-Besse accident on September 24, 1977, a missing electrical relay was to blame; at Arkansas Unit 1 on an unspecified date in 1974, improper venting was the cause; at Oconee Unit 3 on June 13, 1975, a relief valve stuck open because valve components had warped as a result of heat expansion, chemical contaminants had built up on a valve lever, and a bracket in the valve's electrical control system had bent; and at Three Mile Island Unit 2 on March 29, 1978—one year before the major accident—an electrical-system malfunction (a blown fuse) caused the valve to open and remain open. Given the complexity of the valves and their control systems, it is hardly surprising that they are susceptible to many possible types of failure. It is therefore obvious that systematic testing, inspection, and maintenance are essential if this equipment is to function reliably. The failure rate indicates, however, that attention to such tasks has not been a priority of either the N.R.C. or the utility companies involved. The N.R.C. does not classify these relief valves as safety equipment, and so does not review their design carefully in its prelicensing safety procedures. Moreover, the N.R.C. has never required that the valves be tested periodically. After the Three Mile Island accident, the agency did a survey of other Babcock & Wilcox plants and found that none of the relief valves at the plants had been tested since their initial installation.

* * *

If the information that accumulated in the N.R.C.'s own files had been reviewed systematically, it would have alerted regulatory authorities to another safety problem common to the nine Babcock & Wilcox reactors—the excessive frequency with which the main feedwater system was malfunctioning. In just the twelve months before the Three Mile Island accident, for example, the main feedwater system at one or another of the nine plants broke down on twenty-seven occasions. In fact, the failure of the main feedwater system—which left the reactors without their normal method of heat removal—was responsible for many, though not all, of the relief-valve openings in these plants. (The reason was that a main-feedwater-system failure, compounded by other design weaknesses, such as the time delay in the reactor-shutdown controls, produced a sudden pressure surge in the reactor.)

The causes of the failures in the main feedwater systems, like the causes of malfunctions in other complex nuclear-plant systems, were varied. Human error, of course, both in the initial installation of the main feedwater systems and in their subsequent operation, played a prominent role. At Three Mile Island on November 3, 1978, for example, an instrument technician working in the basement of Unit 2, near the polishers, accidentally shut off an electrical circuit on the polisher control panel, and so caused all the outlet valves on the polishers to slam shut and immediately stop the main feedwater flow—the same sort of event that brought about the accident on March 28, 1979. On this earlier occasion, the technician mistakenly thought he was turning on a light switch.

Despite Creswell's concern, the N.R.C. did not begin collecting information about the frequency with which the main feedwater system at Babcock & Wilcox plants malfunctioned. This problem was beyond the scope of the safety aspects that the N.R.C. customarily dealt with. Like the relief valves, the main feedwater system was classified as "non-safety-related" equipment, and since the commission had not officially concerned itself with the safety aspects of the main feedwater equipment in the first place, it allowed the individual utility companies to design their main feedwater systems as they saw fit. The utilities, in turn, delegated this task to their own designers or architect-engineering consultants. As a result, there is no standard design for the main feedwater system in plants using Babcock & Wilcox reactors.

The N.R.C.'s official justification for exempting the main feedwater system from safety review was its requirement that plants have redundant emergency feedwater systems, the theory being that these would eliminate any safety problem that could otherwise be caused by a failure in the main feedwater system. However, emergency feedwater systems have not always been part of the safety equipment required for pressurized-water reactors. The importance of emergency systems was not fully appreciated until about 1972, when an anonymous letter to the A.E.C. led to the recognition of the role that this equipment would have to play in controlling certain types of accidents. Although the N.R.C. mandated the installation of such systems in all new nuclear-power plants, it chose not to apply this decision to plants already operating or under construction. The three Oconee plants, for example, which used the first large Babcock & Wilcox reactors, operated until mid-1979 with only a single emergency feedwater pump each, rather than with the duplicate pumps now called for. Moreover, apart from requiring that some type of equipment be installed to provide emergency feedwater, the N.R.C. has left unanswered a variety of questions about how the equipment is to be designed, controlled, and powered to ensure its reliability. These unanswered questions have been allowed to remain on the N.R.C.'s official list of generic "unresolved safety issues" for several years—a fact that has drawn repeated protests from a number of the commission's own safety reviewers.

If the N.R.C. had checked on the performance of the emergency feedwater systems at the Babcock & Wilcox reactors, it would have found them to be remarkably unreliable. According to the official records, the emergency feedwater systems at the nine plants had malfunctioned at least fifty-three times before the Three Mile Island accident. (Given the somewhat informal official record-keeping for such mishaps, the actual figure may be much higher.) These malfunctions included numerous occasions when the emergency feedwater pumps failed to start as well as occasions when, having started, they broke down. Other cases involved various deficiencies that could degrade the performance of the system under accident conditions. Among the reasons for these failures were operator errors, defective electrical switches, jammed valves, improperly closed valves, loose nuts, insufficient lubrication, corrosion, electrical short circuits, and, in two cases, a component that malfunctioned because it had been bumped or stepped on. Although the Nuclear Regulatory Commission had received numerous reports of malfunctioning emergency feedwater systems at Babcock & Wilcox plants, it failed to oversee corrective action. All nine plants using Babcock & Wilcox reactors, which in certain accident circumstances would have to depend on these faulty emergency feedwater systems, were allowed to continue operating.

The N.R.C.'s general failure to alert the operators of Three Mile Island Unit 2 to some of the known or suspected deficiencies in the equipment installed there—and to the findings of Creswell, Israel, and Michelson, in particular—has been cited in an unprecedented damage claim that General Public Utilities, the owner of the plant, recently made against the agency. The company maintains that the N.R.C. "knew or should have known" from its investigation of incidents such as the one at Davis-Besse in September 1977 that Unit 2 faced serious, uncorrected safety problems, but that the N.R.C. "negligently failed to warn" the company and allowed the plant to remain in operation. The damage claim seeks four billion dollars from the N.R.C. to cover the cleanup operations—which threaten the company with bankruptcy—as well as other losses and expenses it has incurred or expects to incur as a result of the accident. Some observers believe that the claim—filed on the standard government claim form that might be used, say, by someone who slipped on an icy post-office sidewalk—is merely a ploy in the company's lobbying effort to get federal funds for the cleanup work. Lawyers for the company insist, however, that the damage claim is not frivolous and that it will be pursued through the courts. The N.R.C. has rejected the G.P.U. claim, and senior agency officials have privately expressed amazement that in an industry that routinely protests government overregulation a company is now arguing that it was not regulated enough and should not have been allowed to operate its plant. As one N.R.C. commissioner remarked, "It seems like the new industry position is a plea to 'Stop us before we kill again.' "

# 3: A Special Red Light

"During the hour before, I was just doing the normal checks that we do on our shift, just monitoring meters," Ed Frederick—one of the two operators on duty in the control room when the Three Mile Island accident began—has recounted.

"I wasn't aware of anything unusual," his partner, Craig Faust, remembers, adding,

> The next step came when I was walking over to my desk—mine faces the control-room panel. Ed at this time was standing with his back to me. He was turning to look at me, and we were shooting the breeze about something, when I caught the first alarms coming in. Now, when I say "caught the first alarms coming in," I was far enough away that's all I could say. I pointed and at the same time said, "We're in trouble! Something's gone wrong in the plant!"

What Faust and Frederick were witnessing was the sudden flashing of dozens of control-room alarm lights, devices that instantaneously reported the major breakdown of the plant that occurred just after 4 a.m. on March 28, 1979. A wailing siren, which was also connected to the automatic alarm system, reinforced the message. Faust and Frederick were experienced reactor operators, and each of them had been through emergency reactor shutdowns several times before in the years he had spent in the nuclear program. Their training and experience had familiarized them with the established procedures they were expected to follow to secure the reactor, and they did not panic when confronted by rows of glowing alarm lights. According to the official postaccident investigations, they were alert and at their proper duty stations when the accident began, and they immediately executed the preestablished emergency-response plan. They treated the event as a routine emergency—an unscheduled shutdown of the type that any power plant will experience several times a year.

Faust and Frederick had never been forewarned about the peculiar cooling-system problem that can disable, and possibly render uncontrollable, the nuclear-power reactors designed by Babcock & Wilcox, and they did not recognize this problem when it materialized on March 28, 1979. They had not been told about the earlier related problems affecting other Babcock & Wilcox reactors—such as the Davis-Besse incident of September 24, 1977—nor had they been given any special instructions for handling such a contingency. However "predictable" the accident at Three Mile Island Unit 2 (TMI-2) may now appear to have been, given its belatedly documented "precursors," for the operators on duty at the time, the event was a singular and bizarre phenomenon. They plainly knew, as the hours passed, that conditions in the plant were grossly abnormal—"weird" and "all goofed up" are terms they used to describe them—but they didn't know why. Unable to diagnose the key safety problems facing the plant, they were unable to keep the reactor under control. Plant officials had so little understanding of the rapidly developing crisis underlying the strange behavior of the TMI-2 reactor that they waited for more than three hours before declaring a General Emergency and calling for outside help.

The malfunction that complicated what would otherwise have been a simple, unscheduled shutdown of TMI-2 could easily have been prevented from developing into a major accident. The principal problem that affected the safety of the plant was the continuous leak of cooling water out of the reactor as a result of a pressure-relief valve that had stuck open. Another valve, called a block valve, was installed at TMI-2 for the express purpose of cutting off the flow of cooling water from the reactor to the relief valve. All the control-room operators had to do to stop the leak was to close the block valve, and this could have been done with a single switch that was located on a central control-room console, right next to the other controls for the relief valve itself. Faust and Frederick did not execute the simple maneuver that could have swiftly terminated the accident. Although they knew the purpose of the block valve, they didn't know it should be closed, since they were unaware that the relief valve was stuck open.

There is a special red light in the TMI-2 control room that is supposed to help the operators tell when the relief valve is open. Shift supervisor William Zewe, who rushed into the control room seconds after the accident began, to assist Faust and Frederick, has called it "probably the brightest light on our entire console." The light is prominently positioned on the central control console directly in front of the operators' desk. Control-room personnel say that they checked this light many times during the critical early phase of the accident and found that it wasn't on, a fact that led them to conclude that the pressure-relief valve was closed—as it was supposed to be.

Although not included when the TMI-2 control room was initially designed and built, this red-light indicator was installed as a result of a minor mishap that had occurred one year earlier, on March 29, 1978, the day after the TMI-2 reactor was first turned on for preliminary testing. This event, which had no serious consequences, involved a blown fuse that caused a temporary power failure in the electrical control system for the relief valve. As a result of this difficulty, the relief valve stuck open. The plant was shut down, the problem in the electrical system was duly corrected, and Metropolitan Edison officials decided to install an indicator that would help the operators to recognize when the relief valve was stuck open. During this episode, they had been unaware of the fact that the relief valve had stuck open, and company officials didn't want this mistake to be repeated. Since the special red light was put on as an afterthought, this light, unlike other indicators and control devices on the control-room console, didn't have a metal placard beneath it identifying it or describing its function. Instead, the operators glued on a homemade plastic-tape label below the light, stating that "light on" meant that the relief valve "RC RV 2" was "open."

Despite its importance, the special red light was not designed properly and was inherently incapable of providing reliable information about the position of the valve. (The engineering staff at the plant had complained about the design, but plant management—in a March 16, 1979, memo—had rejected the recommended improvements as "not necessary.") For one thing, if the light was *not* on, the operator had no way of telling whether this was so because the relief valve was closed, or simply because the light bulb had burned out, since no test switch was provided that the operators could flip to check whether the bulb was working. Beyond this minor problem, though, was a more basic flaw in the design: the special red light was not actually connected to the relief valve but was merely attached to the automated control system that operates the relief valve and provides the electrical impulses that cause it to open or close. What this light really reported, then, was the intent of the plant's electrical-control systems to open or close the relief valve, and not the true position of the valve. (This scheme resembles the type of distorted battlefield reports that generals sometimes complain about: it relayed only the good news from the front.) The use of "indirect" valve-position indicators is not a unique feature of TMI-2; it is a common valve-instrumentation arrangement found in other U.S. nuclear-power plants. Babcock & Wilcox, according to company Vice President John MacMillan, now agrees that "some more positive method of identifying the position of that valve" is necessary.

* * *

Nuclear-power-plant operators, like pilots flying "on instruments" in the fog and rain, are completely dependent on the essential correctness of their control-panel instrument readings. The pilots, but for the instruments, cannot tell where the airplane is nor can they see the lurking mountaintop that may be in their paths; they need altimeters and navigational equipment to advise them that they are high enough and on the right course. Neither can unaided control-room operators tell what the conditions are inside an electric-generating station's nuclear reactor—the sealed steel cauldron that holds the uranium fuel. Though only a few dozen yards from the control room, the reactor is shrouded by thick layers of concrete shielding to protect the operators (and the neighboring population) from its penetrating nuclear radiation. To monitor the reactor, and other key pieces of equipment used to control it, the operators must rely on thousands of electronic probes and sensors extending out through the plant from the control room. Reports from these devices, which serve as stethoscopes that continuously check the plant's vital signs, are displayed by the meters, gauges, and indicator lights spread out before the operators in the control room. During normal operation, the operators use the data from this electronic intelligence-gathering system to verify that all equipment is performing satisfactorily; during emergencies, they use the same system to find out what equipment is *not* performing satisfactorily. Only after a correct diagnosis of plant conditions can the operators, by remote control, direct the remedial action needed to terminate an accident. Without reliable control-room instruments, the operators could easily be outmaneuvered by mischievous developments that they could not discern. In effect, they would be "flying blind."

A number of obvious factors must be considered in the design of control-room instruments to make sure that the operators always get the critical data they need. The general philosophy on nuclear-plant-instrument design is stated in the standard reference work on the subject by Stephen Hanauer and Clinton Walker:

> The designer who recognizes the role of the operator will give him easily comprehended displays to make the correct interpretation of the [instrument] readings routine. Instrument systems whose displays keep secrets from the operator invite misinterpretation of the readings they do provide, misoperation, and eventual inclusion in accident reports. It is especially important that the designer foresee the operator's need for information under unusual circumstances, including accident and post-accident situations. The operator should have available, possibly in a computer memory, complete information regarding the status of his protection

system. Any components known not to be in working order should be so indicated in an obvious manner.

Despite the recognized importance of proper control-room instrumentation, the A.E.C. and the N.R.C. have not conscientiously regulated these aspects of nuclear-power-plant design. Apart from generalized, hortatory statements on these topics, such as the commonsense observations of Hanauer and Walker, federal regulatory authorities have never issued specific and definitive guidelines governing nuclear-plant control rooms. There are, in fact, no more than a few sentences in the federal nuclear-safety regulations pertaining to the control room in the form of "General Design Criteria" for nuclear plants that were added as an appendix to the A.E.C. regulations in 1969. The existing regulations do not specify in any meaningful detail what equipment should be in the control room; how the control room should be set up to provide the operator with access to needed controls and indicators; what specific data should be displayed on the control panels and where they should be located; or any other vital aspects that affect plant safety. (The lack of N.R.C. requirements governing the control room contrasts markedly with the Federal Aviation Administration's specific requirements governing the mandatory aircraft instruments and controls.) Thus, although there are requirements for a multitude of nuclear-plant-safety systems, the N.R.C. has no particular requirements for how the control room that operates and monitors these devices should be designed and built.

Some of the gaps in the formal N.R.C. safety regulations are filled in by the "Regulatory Guides" that are issued from time to time by the N.R.C. staff. Several of these guides mention the control room and/or control-room instrumentation, at least in passing, but they provide few practical specifications to aid designers. One example is Regulatory Guide 1.97, which addresses the important topic of "Instrumentation for Light-Water-Cooled Nuclear Power Plants to Assess Plant Conditions During and Following Accidents." The basic notion behind Regulatory Guide 1.97, according to a special N.R.C. Instrumentation Task Force that has reviewed it, is that the operator needs to have "appropriate" information in order to be "effective" in carrying out emergency actions to cope with serious accidents. "The operator must be able to identify the nature of the accident," the minutes of one Instrumentation Task Force meeting explains:

> Knowing what systems are required, he must then be able to evaluate the performance of these protection and emergency systems. His ability to then follow the course of the accident and determine if conditions are degrading . . . will allow him to determine the need for operator action to protect public safety. . . . Post-accident instrumentation will assist the operator in making intelligent and accurate decisions for manual actions. Such intervention can correct a deteriorating situation or improve the plant's capability to maintain a safe condition.

Regulatory Guide 1.97, the N.R.C. Instrumentation Task Force has determined, although well-intentioned, is far too vague to be of any real help in assuring that designers will provide proper control-room instrumentation. This is so, the agency's experts say, because the guide simply sums up the vague, qualitative, "currently applied acceptance criteria for instruments" but "does not identify" the specific aspects of the plant that need to be monitored during accidents, or the actual instruments that should be installed for this purpose, "or even specific accidents" that the operators should be able to monitor.

Expert advisers have repeatedly urged the N.R.C. to adopt basic rules to ensure that control-room operators will have all the information they need to respond effectively to major accidents. One recommendation that has received particular emphasis has been to install in each control room a simple instrument that will tell the operators of pressurized-water reactors how much cooling water is in the reactor. After all, managing the plant's cooling water, and seeing to it that there is always enough in the reactor to keep the temperature of the uranium fuel under control, is the operator's most critical responsibility. The N.R.C., nevertheless, has hesitated to promulgate detailed instrumentation requirements or other rules for control-room design because its long-standing policy has been to treat the control room as a "non-safety-related" part of each nuclear plant. The prevailing A.E.C. and N.R.C. "philosophy" on important safety aspects of control-room design, according to a recent N.R.C.-sponsored study on the subject, has been to "leave it up to the utilities." Consistent with this approach, the federal "Safety Evaluation Report" for TMI-2, the official basis on which the plant was issued a federal construction permit, does not even mention the plant's control room.

* * *

Since there are no fixed standards governing their design, no two nuclear-plant control rooms are exactly alike: each is a makeshift arrangement devised according to the preferences of whatever company happens to be building the plant. The large control room at TMI-2 is set up as a kind of amphitheater with the operators placed in the middle of two large U-shaped control panels, one nestled inside the other, separated by a narrow aisle. The lower panel closest to the operator is referred to as the "main panel"; the other, higher one is called the "back panel." There are also two shorter straight panels on the sides of the room, concealed behind the back panel. The four major panels are divided into forty subpanels that are densely packed with tiers of pushbuttons, select switches, toggle switches, dials, gauges, data-recording equipment, colored lights, and, especially along the top of the back panel, row after row of alarm lights. There are, altogether, more than six thousand separate components in the TMI-2 control room, spread out over about nine hundred square feet of panel space.

Metropolitan Edison, as the owner of TMI-2, was delegated the final responsibility for the design of the control room. Like other electric utility companies that build nuclear plants, Metropolitan Edison passed this job on to one of the large architect-engineering firms that serve the power industry, in this case, Burns and Roe. Burns and Roe control-room designers, at least in principle, had to weigh a variety of options in the organization and selection of the controls and instruments to be installed at TMI-2. There are advantages and disadvantages—"trade-offs," the engineers call them—to each of the various possible approaches, and often there are conflicting considerations. For example, one might want to use instruments with large dials, which can be read easily by the operators at a distance of several feet; but a control room using a great many of these devices might be so big that too much operator time would be spent moving from one end of the room to the other.

Cost, of course, is always one of the designer's major considerations. The obvious financial issues involve far more than a cost-accountant's narrow questions about the relative prices and advantages of one specific type of control-room component versus another. The overall control-room "design concept"—the master plan selected by the designers—will be the dominant factor influencing its cost. A heavily automated control room full of sophisticated computers, for example, is obviously more expensive to build than one in which the operators, unaided by this equipment, are left to control the plant. Similarly, a streamlined control room, which makes use of advanced information-processing equipment to help diagnose plant conditions during emergencies, can also be very expensive to build. The markedly different control-room design concepts, accordingly, give rise to very substantial trade-offs between safety and cost: in a sudden crisis, the more advanced, highly automated control room could greatly enhance the effectiveness of the operators' emergency response, but such a control room is initially far more costly to design and install. There are no federal safety rules for control-room design that specify how much of this high-cost, high-technology control and instrumentation equipment must be installed. It is left to each control-room designer to make ad hoc decisions on how best to balance public-safety considerations against the financial costs to the client utility company.

In the case of the TMI-2 control room, the weighing of trade-offs between safety and cost was done only casually, if at all. Since nuclear-plant designers are acutely sensitive to costs, they almost always peremptorily exclude additional safety apparatus beyond the bare minimum required by the N.R.C. (They justify this on the ground that N.R.C. requirements are already excessively conservative, making further safety "extras" unnecessary.) Accordingly, the TMI-2 control room was not designed from a safety point of view at all. Unhampered by the N.R.C. specifications, Burns and Roe considered their assignment from Metropolitan Edison to be the design of a no-frills-added control room suitable for controlling the plant during normal operation. The emergency functions of the control room were treated as peripheral considerations, and no special features for this purpose were provided—such as instruments that would allow the operators to monitor reactor water level or core temperature during the course of an accident. No systematic studies were carried out by Burns and Roe on how the TMI-2 control room would perform during emergencies, nor did the designers carefully evaluate how the layout of the controls and instruments might help or hinder the operators in their attempts to control serious accidents. So little attention was given to the crisis-management role of the operators that Burns and Roe engineers, when they originally selected the basic layout of the TMI-2 control room, didn't even know how many operators would be on duty when the reactor was running.

* * *

The central planning scheme for the TMI-2 control room illustrates its designers' priorities. In a prime location—in the center of the inner U-shaped control console—are the controls most frequently used to facilitate the power plant's everyday power-producing operations. These controls are placed within easy reach directly in front of the operators—a major convenience, given the large size of the control room. Emergency information displays and controls, in contrast, are not so advantageously situated. There is, in fact, no central "emergency station" in the TMI-2 control room—no single, easily accessible panel or location that incorporates the most essential instruments and controls that might be needed during serious accidents. Quite the contrary, emergency controls and information displays, according to a recent study done for the N.R.C., are spread out "in a seemingly random fashion" along the long, densely covered back panel. Its meters and other information displays are so small and far away from the operators that they cannot easily read them. Many components of the emergency controls are relegated to extreme positions on the control panels, at greatest remove from the operators' customary station in front of the main panel. Some of the instruments that the operators might need during serious accidents were even put on the back side of the control panels, so that the operators cannot see them at all from their central command post.

Because the TMI-2 control room is so large, an operator might have to walk as much as fifty feet from his normal station to gain access to some of the emergency equipment. One study of current nuclear-plant control rooms investigated the number of different parts of the control room to which an operator would have to go in the event of an accident requiring the operation of the plant's emergency cooling system. It found that monitoring and verifying the performance of the emergency cooling apparatus "would require evaluation of information from most of the panels in the control center." Other recent studies, focusing specifically on the TMI-2 control room, found numerous deficiencies in the layout of controls and instruments that make it difficult for the operators to do their job.

The difficulties posed by the diffuse and illogical arrangement of emergency controls over nine hundred square feet of control panels at TMI-2 are all the greater because of the small number of control-room operators on duty on any given shift. The N.R.C. required just three licensed operators per shift, only one of whom actually had to be present in the control room on a continuous basis. In the opening phase of the TMI-2 accident, Craig Faust described himself as "sort of bouncing" from station to station in the control room trying to find out just what was happening to the plant. Having to run around the control room can be fatiguing and stressful, hardly conducive to the operators' making an orderly appraisal of plant conditions and selecting the proper course of action to deal with an accident.

* * *

A sprawling, chaotically organized control room, in addition to hampering the operators physically, confuses and distracts them by providing an excessive, overburdening amount of information during an emergency, more than they conceivably need or can make use of. Standard control rooms, like the one at TMI-2, have thousands of signal lights and information displays. A serious accident—or even just a sudden shutdown due to some mundane cause—immediately triggers a near-simultaneous outpouring of signals and alarms from virtually all parts of the control room. The control-room panels are sprinkled with arrays of red, green, blue, white, and amber lights. As the operators often describe it, the panels "look like a Christmas tree" during any emergency shutdown. In theory, the control room's signal lights and instrument displays provide the operators with all the information needed for a correct accident diagnosis and a proper emergency response. In practice, the cluttered control panels deluge the operators during emergencies with a superabundance of information that gravely overtaxes their analytical capabilities. Understanding what is happening to the plant becomes exceedingly difficult because, instead of presenting the operators with the key symptoms, the control-room instruments and signal lights force them to ponder thousands of irrelevant clues. Correct diagnosis of serious accidents, which should be made as straightforward as possible, more commonly requires the operators to solve an unprecedentedly complex technological mystery story.

The level of electronic babble that fills current nuclear-plant control rooms during serious accidents has been studied recently by Thomas Sheridan, a professor of engineering and applied psychology at the Massachusetts Institute of Technology who heads the university's Man-Machine Systems Laboratory and also serves as consultant to the nuclear industry. According to Professor Sheridan, in just the first minute of a loss-of-coolant accident, some five hundred signal lights in a nuclear-plant control room will go on or off, and the following minute, another eight hundred will be activated. "Clearly, this is far more than the operators can cope with for minute-by-minute diagnosis during an emergency," Sheridan comments.

The operator's difficulties in sorting out the information received in the control room during serious accidents would be minimized if sophisticated computers or information-processing equipment were installed in the control room to help with the diagnosis of plant conditions. No such useful apparatus was added to the TMI-2 control room. "Conventional control-room design philosophy," as one designer explains it, "emphasizes providing the *operator* with basic raw data and relying on him to mentally process the data to determine optimum plant operation." Thus, nuclear-plant control-room designers, instead of aiming to simplify the control room by using advanced technology, have until recently "improved" control-room designs simply by adding more and more controls and information displays. They have done so without regard to the fact that they long ago exceeded the saturation point of the operators. Over the last decade, the number of devices in nuclear-plant control rooms has increased from three thousand to around seven thousand, according to a 1977 study on control-room design done by the Aerospace Corporation for the N.R.C. This report concluded that the operators' effectiveness in controlling serious accidents was greatly undermined by gigantic, poorly organized control rooms that "thrust an immense amount of superfluous data" on the operators in critical emergency situations.

Such an information overload occurred immediately after the accident began at TMI-2. "I saw lots of alarms," Ed Frederick recalls, adding that because so much was happening, there wasn't "the time to figure out what caused it." By the time—two hours and twenty minutes (and many hundreds of additional alarms) later—that someone did figure out the problem, the reactor had already gone out of control.

* * *

In its 1977 report to the N.R.C., the Aerospace Corporation recommended a number of practical steps to improve standard nuclear-plant control-room designs. One of the foremost recommendations called for the application of basic "human engineering" principles. Human engineering is an interdisciplinary approach used extensively in the military and aerospace industries to design control systems for sophisticated equipment, such as jet aircraft and advanced weapons systems. The goal of this work has been to incorporate behavioral considerations into control-equipment design in order to maximize the efficient working relationship between man and machine and to minimize the opportunities for (and the consequences of) human errors. A number of general human-engineering guidelines have evolved that suggest steps for control-system designers to follow in order to provide simple, accessible, unambiguous data to the human operators. Among the considerations are how the controls must be designed and organized to facilitate timely operator actions and to prevent mistakes; and how the operator's limitations—in vision, hearing, dexterity, strength, concentration, memory, and analytical ability—have to be factored into the design of the control systems that an operator is expected to manage. Elementary human-engineering considerations, the Aerospace Corporation study showed, were virtually nonexistent in current U.S. nuclear-plant control rooms.

The N.R.C. took no official action in response to the Aerospace Corporation study recommending that human-engineering principles be applied to control-room design. The issue, like many other previously recognized "generic safety problems," was left "open" for "future resolution." The N.R.C. simply continued to issue construction permits and operating licenses for plants, such as TMI-2, whose control rooms were designed without benefit of human-engineering considerations. Burns and Roe limited their attention to designing the hardware that would go into the control room and neglected even some of the most elementary human-engineering aspects of control-room design. According to a new study by the Essex Corporation, a firm that specializes in human engineering, some twenty-six percent of the two thousand information displays in the TMI-2 control room could not be seen by a very short operator standing in the operator's normal position. An operator who happened to be taller would do better, in general, but even a taller operator would not be able to see all sixteen rows of lights on the indicator panel that report on malfunctioning in the plant's emergency safety equipment. "Due to placement and organization of this panel, a 6-foot operator can see only 8 rows of lights from his normal operating position," the Essex Corporation analysts determined.

Seven hundred and fifty alarm lights were installed in the TMI-2 control room, principally along the top of the back panel, facing the operators, to help identify any abnormal conditions that might arise in the plant. Some report conditions that require emergency actions, while others represent lower-priority advisories relating to unusual, but not necessarily dangerous, conditions. Almost a year before the TMI-2 accident, operator Ed Frederick, in a note he sent to his superiors at Metropolitan Edison, complained that the alarms were often confusing:

> The alarm system in the control room is so poorly designed that it contributes little in the analysis of a casualty [plant breakdown]. The other operators and myself have several suggestions on how to improve our alarm system—perhaps we can discuss them sometime—preferably before the system as it is causes severe problems.

Frederick's note was filed away, the meeting it requested was never held, and the basic TMI-2 alarm system was left as initially designed.

Resembling the "idiot lights" on automobile dashboards that are illuminated when the oil gets too low or a door is ajar, the individual alarm light is a small white glass "window," five by three inches in size, with a printed message on its face, such as "pressurizer heater ground fault," "reactor coolant outlet temp hi," "reactor coolant pump vibration hi," or "pressurizer level lo." The alarm lights are grouped together in "annunciator" boxes, each of which contains a total of forty-eight alarm lights arranged in six rows and eight columns that are installed along the top of the back panel. When a given alarm lights up in response to an automatic signal from the instruments that detect abnormalities in the plant, it will flash on and off until the operator pushes a button to "acknowledge" the alarm. The light will then stop flashing, but it will remain illuminated as long as the abnormal condition it is reporting persists.

From the standpoint of optimal human engineering, the TMI-2 alarm lights leave much to be desired. The large number of alarms results in an information overload during serious accidents, especially since so many of them are programmed to light up whenever the reactor suddenly shuts down. The alarms are so spread out around the control room and the printed messages on them are so small that the operators can read only a few of them from any given position in the control room. Moreover, the alarm lights that report malfunctions in a given safety system are not necessarily placed in front of the controls for that system; an operator, spotting the alarm and alerted to some trouble, may have to run across the room to correct the problem. Nor are the alarms grouped together in any way that would separate high-priority safety problems from lesser abnormalities. Thus, the alarm lights in the TMI-2 control room reporting a loss of vital cooling water from the reactor are mixed in on the same panel with the alarm lights that report "reactor building elevator trouble" and "turbine building elevator trouble."

Color coding is a simple human-engineering technique that could make it easier for the operators to differentiate between important and less important alarm lights—the important ones might be red, for example. This is not done at TMI-2, where all the alarm lights are white. There is, however, a color-coding scheme for other lights in the control room that, from the point of view of accepted human-engineering standards, does not make much sense. The TMI-2 control room has thousands of small "indicator lights" that report what equipment is on or off, which valves are open or closed, and so forth. The basic color scheme uses red and green: red for equipment that is running and for valves that are open, and green for equipment not running and for valves that are closed. There are several problems with this arrangement. During routine operation, certain equipment is supposed to be running and certain valves are required to be open, while other equipment is on standby and other valves are closed. Thus, during normal operation, the control panels will have a generous assortment of both red and green lights. A valve or piece of equipment that happens to be in the wrong position or condition, however, is not indicated by this color code and would not be obvious to an operator who is scanning the indicator lights. In the event of an accident, automatic signals will open and close valves, turn equipment on or off, and so forth. But, while watching the red and green lights changing, the operators will still not be alerted by the color-coded signals to any equipment malfunctions that may be occurring. Once again, as in normal operations, all they will see at a quick glance down the control panels is a Christmas-tree display of colored lights.

Other deficiencies in the TMI-2 color-coding arrangement further compromise its usefulness. Ideally, color-coding schemes work best when each color has a unique, standard meaning, so that operators familiar with the scheme can immediately grasp the significance of any colored signal light. At TMI-2, however, the interpretation of the indicator lights involves troublesome ambiguities. For example, when applied to a valve, red means that it is open; yet, when indicating the status of an electrical circuit, red means that it is closed. Although there is a basic red-green color scheme in effect, each color can have several different meanings when used on different instrument and control panels. Altogether, there are fourteen different meanings for red and eleven different meanings for green in the TMI-2 control room.

On top of the confusions that arise from the ambiguous meaning of any given color, the indicator lights themselves are not all placed in a consistent position on the control panels in relation to the equipment controls. Some of the indicator lights for valve positions, for example, are placed above the valve controls, others below or at either side. It is therefore easy for the operators to interpret the data from the indicator lights mistakenly or to overlook important information.

One helpful way to apply color coding to nuclear-plant control rooms would be to use different colors on key meters and gauges to mark potential danger points. This has been done successfully, for example, by airplane designers. The standard airspeed indicators have a simple color code that warns the pilots when the plane is flying too fast or too slow, tells them when the plane is flying slow enough to let down the flaps or the landing gear, and so forth. The pilot need not memorize the danger points or pause to carry out a set of calculations to determine whether the plane is at a safe speed. No such color coding of key meters and gauges in the TMI-2 control room warned the operators when dangerous thresholds had been reached.

Other major human-engineering deficiencies in the TMI-2 control room, found in studies by the Essex Corporation and others as part of the official investigations of the accident, include:

- Many unnecessarily large controls were installed—such as large "J-handle" (or "pistol-grip") switches that fill an operator's hand. (Such equipment is a carry-over from the controls used in older power plants, where large controls were necessary because the switches on the control panel actually opened and closed valves located behind the panels. At TMI-2, however, the switches merely operate tiny electrical relays.) Valuable control panel space is wasted—and other controls are put out of the operators' reach—by the failure to scale down control size.
- Meters and information displays that report critical data—such as the pressurizer level—are small and hard to read, while noncritical data are put on large, prominently positioned instruments.
- Important switches are not carefully designed to be distinguishable by shape, which would help to ensure that the operators reliably take hold of the right controls. (Airplane cockpits, for example, usually use a traditional shape for the main control that operates the landing gear—it is shaped like a small tire—to help make sure that pilots do not reach for the wrong switch and try to land the airplane with its landing gear still retracted.)
- Instruments on the vertical panels used all around the control room are difficult to read because of the glare of the overhead lights. They also have a parallax problem—that is, many meters can't be read correctly unless the operator is looking at them squarely; at other angles, the instrument needle looks as if it is pointing to a different position on the gauge.
- Adequate emergency lighting to enable the operators to see the instruments in the event of a power failure is not available.

• Critical controls are not well guarded to prevent mistaken operator disruption of the plant. The button that causes the reactor itself to shut down suddenly, for example, is not well protected against inadvertent use. (At a nuclear plant in Virginia, an operator's shirttail caught on one of the controls and disrupted the function of the unit.)

• Key controls are arranged in illogical ways that can contribute to operator errors. Some controls that perform a common function are inconsistently designed. Even neighboring controls that do the same job can have inverted designs: one shuts a valve when turned to the right, the other shuts a valve when turned to the left.

• The control room makes poor use of data-recording equipment such as strip charts (these resemble the moving pen-graphs produced by an electrocardiograph or a lie-detector machine). In one case, a single overloaded strip chart records data on up to seventy-two different aspects of plant performance; the resulting graph—which resembles a Jackson Pollock painting—can be nearly indecipherable.

As part of the Essex Corporation's general survey of the TMI-2 control room, a numerical scorecard was prepared to rate the overall adequacy of key controls. Grades were assigned on the basis of how closely the design of the given controls conformed to standard human-engineering criteria. According to this "human factors evaluation," the TMI-2 controls for vital parts of the plant, such as the reactor cooling system, the pressurizer, and the feedwater system, violated ninety-one percent of the applicable criteria.

* * *

Robert Pollard, a former N.R.C. official, has commented on the respective roles of the operators and of the automated control system in current nuclear plants:

> On the one hand, if the operator is able to diagnose a given set of accident circumstances correctly, and to push the right buttons to bring the accident under control, why isn't this just done automatically? If such an accident can be handled so straightforwardly, the automatic controls should be programmed to do so, thereby eliminating one

more possibility for operator error. On the other hand, if certain types of potential accidents are too hard for the designers to figure out ahead of time how to control, can we honestly expect the operators standing there facing hundreds of flashing alarms to do any better?

The TMI-2 control room does have a computer system, of sorts, that provides some very limited assistance to the operators; but it was not designed for use in an emergency. The computer monitors some three thousand different aspects of the plant. It checks the most important conditions, such as the pressure inside the reactor, every three seconds, and other, less important variables less frequently. It compares each reading with predetermined alarm limits and determines whether an unacceptable condition exists. The computer also keeps a log showing when major equipment in the plant turns on or off, and the operators can use the computer to perform simple calculations. Compared to the current "state of the art" in computer design, the system at TMI-2 is exceedingly primitive, since after collecting raw data about the plant, the computer does nothing sophisticated to analyze it. The TMI-2 computer is simply connected to an electric typewriter, called the alarm printer, which automatically types out a continuous list of whatever abnormal conditions are detected by the computer. This chronicle of abnormalities merely notes the condition, the time at which it was detected, and the time at which the condition returned to normal. The alarm printer is an uncritical recorder of fact—an electronic scribe and not a thinking machine. No interpretive messages are added by the computer to diagnose the cause of any problems that it notes, and no attempt is made by the computer to sort the important from the less important data or to organize them in any logical fashion. The operators are left to make their own ad hoc analysis of the data from the alarm printer, along with all the other bits and pieces of information from the control panels.

The alarm printer provides one advantage for the operators: unlike the flashing alarm lights that come and go on the back panel, the computer at least lists the alarms in sequence and notes the times at which they occurred. Aided by this log, the operators can try to trace the sequence of events and thereby find the possible cause of the reactor's troubles. "As we began to run out of ideas," Ed Frederick noted, describing the operators' attempt to use data from the alarm printer during the accident, "I wanted to review all of the alarms that we received to see if anything was happening that we couldn't see." Unfortunately, the operators could not get the data they wanted from the alarm printer. "The alarms—this is a big problem," Frederick explains. "The alarms come in a hundred at a time, but the computer, the IBM typewriter, just types them out one at a time." The alarm printer at TMI-2 takes 4.2 seconds to print a single alarm message. (The advanced computer printing equipment now available can print hundreds of times faster than this.) Because of the torrent of alarms during the accident, the printer soon fell behind; at one point it was as much as two and a half hours late in its recording of the alarms. (Even during routine shutdowns, TMI-2 operators have noted, the alarm printer is as much as an hour behind in its reporting of plant conditions.) Because it was so far behind the events, the alarm printer contributed almost nothing to the operators' understanding of the accident. If the operators wanted more timely information, they did have a way of making the sluggish alarm printer skip ahead so that it would print out the more up-to-the-minute data that were being collected by the computer, but it would soon fall behind again. In using this procedure to get updated information, moreover, the operators would lose the ability to trace the sequence of events, since all the information backlogged in the computer's memory, waiting to be printed, would be erased when the operators asked the printer to skip ahead. Many valuable data about the TMI-2 accident have been irretrievably lost for this reason.

Another problem that diminished the usefulness of the alarm printer during the accident occurred when the paper in the printer jammed one hour and two minutes after the accident began. This caused a complete breakdown of the alarm printer which persisted for most of the critical early phase of the accident. "I don't know what it was," operator

Craig Faust commented, "but the easiest way for me to say it is, it sort of started eating the paper."

* * *

A review of some of the major events that occurred during the critical early phase of the accident—from 4 a.m. until just after 7 a.m.—indicates how the major design defects in the TMI-2 control room, each in its own fashion, influenced the outcome of the event. The specific defects that impeded the operators' response to the accident were apparent from the opening moments of the event. In the first few minutes, the operators received a vast array of signals from their control panels. Ed Frederick says that one of the first alarms he read said that the turbine-generator had shut down—"tripped," in the operators' jargon—and that he then remembers shift supervisor William Zewe running up behind him and announcing, "We just lost the reactor," meaning that the reactor also had tripped, but the operators had no clear information from their control panels on the cause and had no time to ponder the matter, for as Frederick notes, "We had to react to what was happening." In theory, there is supposed to be at least enough automated equipment in each nuclear plant so that the operators will have several minutes to analyze conditions before they have to act. The so-called ten-minute rule is the informal N.R.C. criterion for the length of the nominal grace period. In the case of TMI-2, the automated controls were programmed to start up the three emergency feedwater pumps whenever the plant suddenly shut down, and did so promptly within a fraction of a second after the main feedwater system failed. However, other operator-executed emergency actions to compensate for the loss of normal methods of cooling the reactor were necessary, and thirteen seconds after the first alarm sounded, Craig Faust lunged for the controls to start an additional pump. Since this additional pump was supposed to be routinely turned on following an emergency shutdown, it should have been operated automatically, yet the control-system designers overlooked this requirement. When Faust tried to start the additional pump, moreover, it wouldn't start. He tried repeatedly, but unsuccessfully, until his partner, Ed Frederick, noted his difficulty and came to his assistance. Frederick succeeded in starting the needed pump, whose control-room actuation switch was improperly designed for emergency use. The problem with the switch was that the operator has to turn it to a start position, hold it there, and then wait for the pump to start. Faust was apparently letting go too soon. In an emergency, when the operator is under great pressure to respond to fast-moving events, an improperly designed switch with a built-in delay is a dangerous hindrance.

While Faust and Frederick were devoting precious time and attention to a problem caused by inadequate automation, two developments of major importance occurred in that opening minute of the accident, which the distracted operators did not notice. The first and more critical development was that the relief valve stuck open. The relief valve opened automatically about three seconds after the accident began to relieve the pressure surge within the reactor (by discharging a negligible amount of cooling water from the reactor). Ten seconds later, the pressure surge was over and the pressure of the cooling water in the reactor was back to normal. The relief valve should have reclosed—but it didn't. The operators remained unaware of this malfunction because the special red light, when they checked it, had turned off. Accordingly, in their attempts over the next few hours to analyze the accident, they ruled out an open relief valve as a cause of the plant's difficulties. There was no other equipment in the control room that directly reported when the relief valve was open and none of the seven hundred and fifty alarm lights contained a warning message that told the position of the valve.

The operators were plainly stymied by the miscue they received from the special red light and by the lack of appropriate instruments and indicators to report relief-valve position. Some analysts of the TMI-2 accident maintain, nevertheless, that the operators still ought to have known—on the basis of indirect evidence—that the relief valve was open. According to this line of reasoning, "operator error" rather than design deficiencies in the plant played the dominant role in the failure to recognize that the relief valve was open. Babcock & Wilcox favors the operator-error theory, and company spokesmen have insisted that there were some conspicuous clues that would have told an alert operator that the valve had jammed. Thus, they point to the data displayed by the control-room instruments concerning the fate of the water discharged by the open relief valve. (This water is fed into a drainpipe that ends up in the reactor-coolant drain tank.) The persistent high temperature in the drainpipe and the increasing water level, temperature, and pressure in the drain tank, they argue, were manifest indicators of an open relief valve discharging the hot cooling water from inside the reactor.

The layout and design of the control room make it obvious why the operators paid no heed to this information, which supposedly told them about the open relief valve: they could not readily see it. The instrument panel and alarm lights showing how much water was accumulating in the drain tank, for example, were on the *back side* of a control panel, far from where the operators normally stand and completely out of their line of sight. It would be impractical for the operators in the middle of an emergency to leave their main position and spend a great deal of time exploring the remote back panels; an N.R.C. Regulatory Guide, in fact, cautions the operators against doing so.

Control-room personnel did, however, make a brief check of the reactor-coolant-drain-tank instruments at least twice during the first fifteen minutes of the accident. Still, they were not alerted to the open relief valve because of the poorly designed drain-tank instruments that did not make this fact at all clear. All the instruments provide is a report on conditions in the drain tank at any moment; there was no recording apparatus to show any changing trends, which an operator, by merely glancing at the indicator, could not determine. The operators had no clear way of knowing whether the amount of water reported to be in the drain tank simply represented the amount in the tank *prior* to the accident, or the amount discharged by a temporary relief-valve opening at the beginning of the accident, or the amount accumulated as the result of a stuck-open relief valve. Only a recording device that showed the trend—the continuing flood of water after the relief valve was supposed to have closed—would have revealed the open relief valve. Since the operators knew that there was water in the drain tank prior to the accident and that the relief valve had opened, at least temporarily, the readings from the drain tank were attributed to these causes.

The temperature in the relief-valve drain pipe—the other possible clue indicating that the relief valve was open—wasn't shown at all by the control-room instruments. To ascertain the temperature, an operator would have to interrupt other duties, go to a computer terminal, and make a special request for the data. The operators did this, as part of a general check of plant conditions during the accident, but did not think that the temperature in the drainpipe was disturbingly higher than normal. After all, they knew the relief valve had opened—briefly, they thought—and that the drain pipe would then take some time to cool down. A further ambiguity affecting the interpretation of the drainpipe temperature data was that the relief valve was known to have been leaking slightly for several months *before* the accident—since October 1978—which made for a persistently higher than normal temperature in the pipe. The high temperature readings they noted during the accident, therefore, did not strike the operators as particularly alarming. The final difficulty affecting their response to the temperature data provided by the computer was that the computer was not actually providing the correct temperature. "My information indicates," John MacMillan, a Babcock & Wilcox vice president, has explained, "that the operators did, in fact, ask the computer what the temperature was . . . and got a reading somewhere around two hundred and eighty degrees Fahrenheit, in that range." Yet the computer, Mr. MacMillan concedes, was only programmed to read the temperature *up to* two hundred and eighty degrees. (Like an automobile speedometer covering a specific range, the computer at TMI had been programmed to report the temperature only within a certain narrow band.) The actual temperature in the pipe, he adds, was probably higher. Since the temperature of the primary coolant, before it was discharged through the relief valve, was in the neighborhood of six hundred degrees Fahrenheit, the operators did not regard the much lower readings they saw as proof, or even as circumstantial evidence, of an open relief valve. "It would be a reasonable conclusion" from such temperature readings, Victor Stello of the N.R.C. explains, "to say that the valve is closed rather than open." The equipment in the TMI-2 control room, despite the general admonitions of Hanauer and Walker, was "keeping secrets" from the operators.

* * *

The second hidden development that occurred in the opening moments of the accident was the complete failure of the emergency feedwater system as a result of two closed valves. The operators, performing in exact conformity to their emergency procedures, had promptly checked right after the shutdown to see that the emergency feedwater system was working properly. Yet because of the defects in their control-room instrumentation, they failed to discover that the equipment was disabled. One of shift supervisor Zewe's first orders, after he announced the reactor and turbine trips over the plant's public-address system some eight seconds into the accident, was to "verify the emergency feed," but a direct order was superfluous because Faust was already rushing to do this. Faust recalled that he "went immediately over" and "looked down and found that all three emergency feed pumps were on." He mistakenly inferred from the fact the pumps were running that they were actually delivering water to the steam generators (where this colder water could help cool the hot water coming from inside the reactor). His mistake arose because there was no instrument on the panel he inspected or elsewhere in the control room that reported the rate at which emergency feedwater was flowing to the two steam generators. In fact, the flow rate was zero because the flow path was completely cut off by the two closed valves. In addition to having no flow meter, the control room also had no alarm to indicate the closure of the two key valves.

It was not until several minutes later that Faust, who was running from one part of the control room to another, returned to recheck the emergency feedwater system. "I came back over and the first indication that something was wrong with feed was that the [steam] generator [water] levels . . . indicated to me that the steam generators were dry," he recalled. Faust started checking to see what might be wrong, which was not an easy task. According to the N.R.C.'s *Reactor Safety Study,* there are at least ninety-eight ways in which a typical pressurized-water reactor's emergency feedwater system could fail to work. The hasty mid-emergency effort that Faust had to make to pinpoint the reasons for this particular failure was further complicated by the poor layout of feedwater controls and indicators in the control room. As the Essex Corporation analysts noted in their review of the TMI-2 control room, "The feedwater panel is not laid out in a sequential or otherwise logical fashion." As Faust initially scanned the randomly organized feedwater panel, he overlooked the lights that indicated the two key valves were closed—two small, undistinguished lights out of the thousands on the "Christmas-tree" control panel. He missed seeing the lights because his body, as he leaned over the horizontal portion of the panel to check other equipment, obscured them. There was also, he found later, a paper tag partly covering one of the lights. This tag, hanging from a nearby switch, was one of the "caution tags" attached to switches and instruments throughout the control room to indicate, for example, that some piece of equipment had a defect or had been taken out of service temporarily for maintenance or repair. TMI-2's Administrative Procedure No. 1037, covering "Caution and Do-Not-Operate (DNO) Tags," did not provide any formal guidance relating to the hanging of the tags to make sure that they did not inadvertently block control-panel instruments or indicator lights. Three Mile Island operators, moreover, are not required by plant procedures to make a routine check of the control panels when they come on shift to verify that critical valves are in their proper position (nor is there any automated double-checking procedure programmed into the plant's control system to do this). Operators in general, and the two on this shift, "assumed" that these two valves were open, according to Ed Frederick, who adds, "and that may be why we didn't know they were shut earlier in the shift. We just started scanning the panel and stuff—you don't notice things—I guess we didn't notice that."

Finally, some eight minutes after the accident began,

Faust found the two closed valves, and shouted his discovery to the shift supervisor. "I was pretty boisterous about it," he recalled. "It was not something I wanted to find or expected to see."

"What do you mean they're shut?" Zewe yelled back.

Faust, who grabbed the controls to open the valves as Zewe and Frederick rushed to his side, had no explanation. Frederick recalls that when Faust turned the switches to open the valves, "he nearly ripped them out of the panel."

The operators, who were greatly perturbed by the discovery that the emergency-feedwater system was inoperative, subsequently thought that the problems with this equipment might be responsible for another condition they noted in the plant: the abnormally low pressure in the reactor. Faust, after opening the valves for the emergency feedwater system, had a great deal of difficulty over the next hour in manually adjusting the rate at which emergency feedwater was flowing into the steam generators. (Regulating the flow requires a complex balancing of factors to achieve the appropriate rate of heat removal from the reactor; ideally, this would have been done by computer-controlled equipment rather than in a trial-and-error fashion by a harried operator.) Since the cold water supplied by the emergency feedwater normally exerts a strong influence on the pressure inside the reactor—in the same way that pouring cold water on a hot pressure cooker will cool it down—Faust thought that the remarkably low pressure in the reactor (down to around a thousand pounds per square inch from a normal twenty-two hundred psi) was probably attributable to the emergency-feedwater-system problem. Actually, the pressure inside the reactor was low and steadily falling because the relief valve was open. Yet not knowing this, and having no other explanation for the phenomenon of low pressure, the operators continued for hours to fuss with the feedwater-flow adjustments. The two closed valves on the emergency feedwater system contributed to what shift supervisor William Zewe terms the "confusion factor" in the control room during the accident. In effect, as Craig Faust observed, the emergency feedwater problem served to "cover up" the real problem in the plant, the open relief valve. This could not have happened if proper instruments had been installed in the control room to monitor the actual conditions in the reactor.

* * *

Fifteen minutes after the accident began, having detected and presumably corrected the major apparent abnormality in the plant, the failed emergency feedwater system, the operators and their supervisors believed that the TMI-2 reactor had "stabilized." The control-room instruments and information displays did not tell them otherwise. They proceeded, accordingly, with routine postshutdown chores, oblivious of the steadily developing safety problem in the plant. The apparent calm felt in the control room was akin, perhaps, to that of a ship's crew after passing from turbulent waters into the placid seas in the eye of a hurricane. However, unlike a ship's officers, who appreciate the nearby perils, the control-room personnel at TMI-2, monitoring inadequate and highly misleading instruments, lacked a sense of foreboding. The pressurizer-water-level indicator, for example, their customary source of information about the adequacy of the reactor's cooling water supply, falsely told them that there was more than enough water in the reactor, so they were prompted to cut back the flow of water from the emergency cooling pumps, equipment that was automatically activated in the opening minutes of the accident. Instead of delivering emergency cooling water to the reactor at a rate of one thousand gallons every minute, as designed, the throttled pumps provided a mere trickle of twenty-five gallons per minute, which was one-tenth the amount leaking out of the reactor through the open relief valve. The operators, moreover, did not simply override the automated emergency-cooling-system controls in just that initial instance, four minutes after the accident began. As the hours passed, there were repeated electronic commands from the automated control system to start the pumps. The operators, relying on the pressurizer-water-level instrument, on each successive occasion cancelled the automatic instructions and prevented the emergency-cooling apparatus from working. They went so far as to lock some of the controls for the emergency cooling system in an "off" position, thereby completely defeating any attempt by the automated control system to start it up. Since they saw no ongoing emergency, they also proceeded to shut off other safety equipment, such as the emergency diesel generators that had started automatically to provide backup power that could be used, if necessary, to operate plant safety systems. Overriding the automated commands that were trying to actuate urgently needed safety apparatus, therefore, was not a momentary "error" made by operators in the confusing opening minutes of the accident; rather, their actions reflected a strategic decision, prompted by misleading control-room instruments, and they consistently followed it hour after hour.

As early as the first few minutes after the accident began, some of the cooling water inside the reactor began to change into steam, a fact that should have prompted the operators to reassess their view that they were dealing with a routine plant shutdown. The reason steam formed in the reactor was that the pressure of the cooling water in the reactor, which was decreasing steadily as a result of the open relief valve, fell below the "saturation" level, the point at which water at a given temperature and pressure boils. The accumulating steam pockets began to interfere with the flow of cooling water through the reactor and, as they grew, threatened to envelop the core in a steam bubble that would deprive the uranium fuel of needed cooling water. The presence of hazardous steam pockets in the reactor, however, was not appreciated by the operators because there was no instrument in the control room that reported when the reactor's cooling water reached "saturation" conditions. Of course, there were control-room instruments that showed the temperature and pressure of the reactor's cooling water. However, to determine whether saturation conditions exist requires either laborious calculations or reference to complicated "steam tables" that tell whether water is a liquid or a gas at given combinations of temperature and pressure. The standard instruments in the control room merely reported the raw data about pressure and temperature and provided no warning when the pressure fell below the dangerous threshold where the cooling water would suddenly flash into steam. Unlike the pilot who can readily see when an airplane "passes over the red line" into a dangerous condition, the operators had no warning of the peril the plant was in.

The optimistic reports they received from their control-room instruments were so convincing that, some fifteen minutes after the accident began, shift supervisor William Zewe, the most senior official on duty at the plant, left the control room. For the next forty-five minutes, he was in the basement of the plant attending to some of the collateral equipment problems that occurred because of the main-feedwater-system failure that initiated the accident. The apparatus absorbing his attention was in the "non-safety-related" part of the plant. The control-room operators, for the rest of the first hour of the accident, were also preoccupied with a range of minor problems. They did make a telephone call around five minutes after the accident began, to inform station manager Gary Miller of the unscheduled TMI-2 shutdown. He was at home preparing to drive to a meeting in New Jersey. The call was *pro forma* notification that Miller routinely received whenever one of the units shut down. He simply acknowledged the report of the shutdown and continued to get ready for his out-of-town trip.

During the second hour after the shutdown, the apparent calm in the control room was broken by a variety of confusing signals from control-room instruments and alarms. According to these reports, the four nine-thousand-horsepower coolant pumps, the primary means for delivering cooling water to the reactor core, were shaking badly. (The pumps are designed to pump only water, and they were choking on all of the steam that was building up in the reactor.) Other signals indicated abnormally high temperatures in the containment building housing the reactor. This was a result of hot water being dumped into the building through the open relief valve. (The water would normally have gone into the drain tank, but the continuous leak caused the drain tank to fill until it burst, and the water then went directly onto the floor of the building.) The operators, however, who had been having continuing difficulties in adjusting the supply of emergency feedwater to the steam generators, thought the water in the containment might be from a leaking steam generator. Since they had no specific control-room instruments or alarms to help them pinpoint the location of the leaks, they were forced to rely on hasty guesswork. Spurious fire alarms were sounded in the plant, and there were even repeated indications from the control-room instruments that the nuclear chain reaction might be restarting in the reactor. The operators had thought the reactor safely shut down because all the control rods had been inserted in the first moments of the accident, yet, more than an hour later, the reactor unaccountably showed new signs of life. The operators were forced to resort to "emergency boration"—that is, they added the chemical boron, which, like the control rods, absorbs neutrons and curtails the fission chain reaction. Of course, the chain reaction wasn't really restarting, or in any danger of doing so. The instruments that monitored the chain-reaction power level were simply sending spurious signals to the control room. (These devices are designed and calibrated to work when the reactor is full of water; now that a frothy mixture of steam and water was present within the reactor, instruments were thrown off kilter and were producing misleading data.) Unaware of the steam and of its effects on the instruments that monitor the core, the operators had to handle the further headache of carrying out the emergency boration.

According to a recent engineering analysis, the loss of cooling water through the relief valve, which was two-and-one-quarter inches in diameter, did not produce a critical cooling-water deficit until almost two hours after the start of the accident. By that point, about half of the cooling water normally present in the reactor had been lost, thus exposing the tops of the twelve-foot-long fuel rods so that the uranium fuel started to overheat drastically. It was estimated that at any point up to about 5:40 a.m., closing the block valve and turning on the emergency cooling pumps could have saved the core from serious damage. Even after 6 a.m., when the destructive overheating of the core had begun, control-room instruments continued to keep the evidence from the operators. Inside the core, according to postaccident analyses, the temperature of the Zircaloy fuel rods—the metal tubes holding the uranium fuel—increased from the normal six hundred degrees to more than four thousand degrees.

The enormously high temperatures that were achieved

had a catastrophic impact on the fuel rods. They swelled and ruptured, and the Zircaloy itself began to react chemically with the hot steam surrounding the rods. The reaction, which has been extensively studied, involves the oxidation of Zircaloy—literally the burning of the Zircaloy metal—at temperatures higher than about fourteen hundred degrees. (At high temperatures Zircaloy has such a strong affinity for oxygen that it rips the oxygen out of water molecules.) The onset of this reaction signals the beginning stage of a possible meltdown accident. At about two thousand degrees, the reaction speeds up and becomes very hard to stop since the reaction itself is exothermic (gives off additional heat that drives the reaction faster). Hence, as the reaction progressed, more and more Zircaloy metal was turned into brittle, crumbling zirconium oxide, and the tubes holding the uranium fuel were destroyed.

The radioactive waste materials that accumulate inside the core during reactor operation normally remain tightly locked inside the uranium fuel pellets. A small fraction of the gaseous radioactive "fission products" escapes from the fuel pellets into the space between the pellets and the Zircaloy tubing. When the tubes, which expanded as a result of the high temperatures, finally burst, these trapped radioactive gases were released into the reactor. Moreover, when the uranium itself, like the Zircaloy, began to overheat, additional radioactive gases and other volatile fission products were driven out of the overheated uranium fuel pellets. TMI-2 accident analysts have referred to the two to three hours after the accident began as the "bakeout period" in which the reactor fuel was "cooked" and forced to release massive amounts of radioactive material into the reactor. The N.R.C. staff estimates that about thirty to forty percent of the zirconium in the reactor was consumed by metal-water reactions during the accident, destroying the central and upper portions of the core. The agency staff, which had previously maintained that an accident capable of damaging even a small part of the core was so unlikely as to be an "incredible event," has had to invent a new term to cover the situation at Three Mile Island. "We now refer to the top part of the core,"

Dr. Roger Mattson said recently, "as having been 'rubble-ized.' "

In the control room, meanwhile, the operators remained completely oblivious of the soaring temperature and ferocious chemical processes that were destroying the reactor's uranium-fueled core. There was, after all, no instrument installed in the control room for measuring core temperatures under accident conditions. Such instrumentation had been recommended for years by leading nuclear-safety authorities, who felt it was vitally important for the operators to know the temperature of the uranium fuel. Yet, N.R.C. did not require plants to have this equipment, and none was provided by Burns and Roe, the control-room designer, or by Babcock & Wilcox, the reactor designer and manufacturer.

Purely by chance, there were some thermocouples—temperature-measuring devices—present in the TMI-2 reactor when the accident occurred. Located about twelve inches above the top of the core, these thermocouples were not standard equipment on Babcock & Wilcox reactors and they were not added for safety reasons. Installed as part of an experimental study of core performance, the thermocouples were a temporary instrumentation feature of the plant, connected to the control-room computer for measuring temperatures during normal operation. Accordingly, if a control-room operator requested temperature data from the computer, he would receive useful information only when the temperature was within the normal six-hundred-degree range. When the temperature got above seven hundred degrees, the computer, instead of reporting it, would simply print out a string of question marks—"???????." Although the thermocouples could actually measure much higher temperatures, the computer was not programmed to pass these higher temperature readings on to the operators. The computer would also print out question marks when the temperature fell below the normal operating range, and when the thermocouple itself malfunctioned. There was an urgent need for timely, reliable data about the temperature in the core in the critical period between 6 a.m. and 7 a.m. on March 28; what was available from the computer was mostly question marks.

By the time the block valve was finally closed at about 6:18 a.m.—and this was done because Brian Mehler, the shift supervisor for the incoming shift, surmised that the relief valve might be open—the uncontrolled overheating of the core was already under way. While closing the block valve temporarily slowed the developing crisis, it was done too late to end it. For after the block valve was shut, the operators still failed to see the need to start up the emergency cooling apparatus. They believed that the core was adequately bathed in the required amounts of cooling water. Not only did they keep the emergency cooling equipment throttled, but they also allowed the separate drain line, opened earlier to relieve the apparent excess of cooling water, to remain open. The loss of coolant continued.

Even when radiation alarms sounded shortly before 7 a.m., triggered by the radioactive material that was escaping from the overheated fuel, the ambiguous data the operators received from the control-room instruments failed to alert them to the fact that severe damage had been done to the reactor core. Thinking that the reactor was filled with water, they had no reason to believe that the fuel could overheat. And unless the fuel overheated, how could it release any large quantity of radioactive materials? The operators, trying to interpret the radiation alarms in this situation, were also stymied by the fact that many radiation instruments had gone "off-scale"; the radiation levels, in other words, were higher than the instruments indicated, but the operators had no way of knowing how much higher they were. The operators later reported that they initially attributed the radiation alarms to some small quantity of radioactive debris that had shaken loose somewhere in the plant, a possible "crud burst," similar to what happens when rust breaks loose in an automobile radiator. In this case, the "crud" they had in mind was the slightly radioactive material that can build up on the outside of the fuel rods.

It wasn't until late the next day that they realized the core was uncovered and badly damaged. (This was confirmed by the sample of cooling water taken late Thursday afternoon, March 29.) At 7:53 a.m. on March 28, when shift supervisor Zewe declared a "site emergency," he was aware that there was "a tremendous problem" in the plant, but he had no real appreciation of the true peril the plant faced. The declaration of a "site emergency"—soon to be upgraded to a "General Emergency" by station manager Gary Miller—was a required action because plant procedures stipulated that this had to be done as a precaution when control-room radiation alarms reported abnormal conditions. On the basis of the data then available in the control room, the operators had only the dimmest understanding of the nature of the actual emergency that was besieging the plant.

It was not until postaccident technical analysis many months later that the full dimensions of the TMI-2 "General Emergency" were properly appreciated. At the moment the block valve was closed, it was estimated that the reactor was within thirty to sixty minutes of a core meltdown. By that time, the N.R.C.'s analysts calculate, "up to approximately one-half of the fuel in the core could have been in liquid form." This "could then have resulted," they add, "in an irreversible core heatup and meltdown."

The impediments in the TMI-2 control room to the diagnosis of serious accidents were so pervasive that the operators were kept totally ignorant of the fact that the reactor was hovering on the brink of the ultimate nuclear accident. They were frustrated and confused at every turn by missing instruments, unreadable meters, ill-placed monitoring devices, improperly designed switches, inaccessible emergency controls, inherently ambiguous signal lights, a useless computer, distracting and irrelevant alarms, and, most especially, perverse miscues about plant conditions. Only with unparalleled good luck could a handful of operators have been expected to overcome the difficulties posed by a control room so designed. Faust, Frederick, and Zewe had no such large measure of good fortune, and they encountered no serendipity as they responded to what they thought initially was a routine, albeit unscheduled, plant shutdown. The only fortuitous thing they had was a little red light that had not been included in the control room's basic design. Their circumstance was all the more ironic, since with only one piece of reliable instrumentation telling them that the relief valve was open, TMI-2 station manager Gary Miller has lamented, the operators could have closed the block valve and quashed the developing emergency. "They would have been just as lucky as they were unlucky," he added.

# 4: First Team

The immediate responsibility for managing a serious nuclear-power-plant accident falls to the small crew of operators who happen to be on duty when the upset occurs. In the predawn hours of March 28, 1979, this burden was assumed at TMI-2 by four federally licensed reactor operators who were assigned to the March 27–28 overnight shift: Craig Faust, Edward Frederick, Frederick Scheimann, and William Zewe. The operators' formal assignment in responding to an accident at a nuclear plant has parallels with an emergency-room doctor's role in treating the victim of a sudden life-threatening illness. In both instances, a very complex system is exhibiting signs of disruption and distress, and the task of providing needed treatment demands special skills and the use of highly sophisticated equipment. In both cases, before a suitable emergency response can be undertaken, the problem must be diagnosed correctly and the proper method of treatment must be selected. A doctor is given years of comprehensive professional training and customarily serves a closely supervised apprenticeship to develop the skills needed to diagnose and treat serious illnesses. The physician is expected to be more than a mere technician, and so must go through a training program that aims at providing a detailed understanding of the human body and the illnesses to which it can be subjected. This basic general knowledge, as well as a set of specific technical skills, is what the doctor draws upon in making a proper diagnosis.

A control-room operator has no comparable professional training, possesses only sparse general knowledge of nuclear-reactor technology, and benefits from what is in several major respects only a superficial "crash course" on key aspects of nuclear-plant safety. Even the "senior" operators, who are the highest-ranking officials legally required to be in the control room when the nuclear plant is operating, are not required to have an advanced engineering or nuclear-sciences degree. The only educational requirement imposed on them specifies a "high school diploma or equivalent." The operator's role may parallel the doctor's, and the operator's responsibilities may even be greater, given the potential scale of injuries that can result from an improperly handled nuclear-plant emergency. Still, nuclear-plant control-room operators in the United States are not officially required to possess the physician's high level of professional training and competence. As with most of the approximately twenty-five hundred federally licensed control-room operators, high-school graduation was the highest educational milestone achieved by the four operators who were in charge of TMI-2 when the accident began.

Alvin Weinberg, the theoretical physicist who contributed heavily to the development of nuclear power in his many years as the head of the A.E.C.'s Oak Ridge National Laboratory, has repeatedly called for the formation of a "special priesthood" to oversee the operation of nuclear plants. Such a select cadre of highly trained and disciplined technical specialists would be able to provide the level of competence he feels is a key factor in the safe operation of large nuclear plants. No one in a position to implement this recommendation has ever paid much attention to Weinberg's advice. The official attitude has been that operating a nuclear plant is a relatively mundane chore that calls for an unexceptional level of technical ability.

The operator's basic job—"just monitoring meters" is the way Ed Frederick describes it—consists of routine surveillance of a plant that, in normal circumstances, largely runs itself. Even during accidents, multiple, supposedly fail-safe emergency systems run by automated control devices are presumed to be capable of controlling the plant without much intervention from the operators. The operator's emergency role is expected to be chiefly one of "monitoring and verifying" the response of automated plant-safety devices to the contingency. The operator, it is reasoned, may have to push a button or two, but will not be required to make decisions that require advanced engineering training.

Operators, accordingly, are not expected to possess keen analytical capabilities or wide basic knowledge. In essence, the control-room operator is looked upon not as a highly skilled professional but as a kind of museum guard who oversees the operation of someone else's technological masterwork. Nuclear-plant operators have an entirely different status from that of a pilot in charge of a Boeing 747, who has passed through an exceedingly thorough screening and a rigorous training program and who receives a salary that may approach one hundred thousand dollars a year. The current salary for a control-room operator, according to an industry newsletter, is about twenty thousand dollars per year, and the annual salary of a senior operator is about twenty-eight thousand dollars. About half of the operators of U.S. nuclear plants, including the four on duty at TMI-2, are former enlisted men in the U.S. Navy who were trained as reactor operators or technicians by the Navy and joined the commercial nuclear program when they left the service; the other half typically began their careers as unskilled or semi-skilled power-plant technicians who advanced to the position of control-room operator.

The nuclear industry has resisted proposals to upgrade substantially the control-room operator's required level of professional qualifications. Industry spokesmen have maintained that higher-skilled and higher-salaried engineers are unsuited to the routine, around-the-clock shift work of operating a nuclear plant, and that they would quickly tire of the job and quit, thereby causing an undesirably high turnover rate in the plant's operating staffs. There is also concern that boredom with the job on the part of "overqualified" operators could lead to a slackening of required discipline and to operator negligence. Indeed, there is already some evidence that operator inattention may be undermining emergency preparedness at current nuclear plants, although to an extent that has not been systematically investigated. For example, Portland General Electric Company officials describe "a significant problem" in maintaining elementary discipline at the company's Trojan Nuclear Power Station in Oregon. According to company reports in March 1979, a control-room operator "had a basketball game piped into his executone phone at his control console" while radioactive water was overflowing a tank in the plant's auxiliary building and flooding the building to a depth of five feet. This incident was not the result of a momentary lapse in discipline. Plant officials noted in internal memorandums that "at the present time the majority of the operators do not give their full attention to operating the plant." The operators' diversions included "reading books, magazines, and newspapers," "sleeping in the shift supervisor's office," "doing crossword puzzles, playing games, tying flies ... drawing pictures and many other non-productive things." The internal memorandums noted further, "All this [is done] with the knowledge of the shift supervisor and, with the exception of one, they do nothing about it and partake of many of the same activities." At another operating nuclear plant—one of the Browns Ferry reactors operated by the T.V.A.—an N.R.C. inspector, visiting the plant a few years ago for a routine inspection, entered the control room and saw no one on duty at that moment, although the reactor was operating at full power. One operator had gone behind the instrument panels to check something. His partner had gone to the men's room.

The possibility that TMI-2 operators may not have been at the ready when the accident began has been investigated by the N.R.C. during the course of its official inquiry into the event. N.R.C. investigators checked to see whether any kind of celebration might have been going on during that early-morning shift, since March 28, 1979, was the first anniversary of the initial startup of the TMI-2 reactor. No such celebration was in progress, the N.R.C. found, and there was no evidence of any type to indicate that the operators on that shift were less than fully conscientious.

What affected the operators' emergency response in this case was not lack of attention but lack of required training. Unlike the emergency-room doctor, who is trained to focus on the patient's "vital signs," or the airplane pilot, who is trained to monitor "danger points" (such as inadequate airspeed, which could cause the plane to "stall" in midair), the TMI-2 operators had little of the basic technical understanding needed to determine whether or not the reactor was in a safe condition. They did not, in fact, have a firm grasp of the basic principles on which a pressurized-water reactor works, and they were unable to determine from the control-room instruments whether the reactor was adequately cooled. This was admittedly not a simple task, given the inadequacy of the control-room instruments and the deluge of distracting and irrelevant information they received during the emergency. Still, there was unmistakable technical evidence in the control room that morning—the readings showing spectacularly low pressure in the reactor—to have shown them that a dire cooling problem was threatening the safety of the reactor. The operators were unaware of the significance of this evidence, however, because they simply did not know that one of the "vital signs" indicating the safety of a pressurized-water reactor happens to be the pressure within the reactor.

A pressurized-water reactor is so named because its proper operation, like that of a pressure-cooker, depends on having the water it contains kept under high pressure. The reactor's cooling water is exceedingly hot because it comes in direct contact with the uranium fuel, and it would otherwise boil away if exceedingly high pressure were not maintained in the reactor. Accordingly, a pressurized-water reactor operates at approximately twenty-two hundred pounds per square inch during normal operation. This is about one hundred and fifty times higher than normal atmospheric pressure, and it is high enough so that the water in the reactor, which is at a temperature of about six hundred degrees, will not boil. The conditions under which water boils and the principles explaining this phenomenon have been investigated by scientists for several hundred years. Thermodynamics, one of the most subtle branches of modern physics, summarizes the accumulated body of knowledge on this subject. The basic facts are easy enough to grasp. Under standard conditions—normal atmospheric pressure at sea level, which happens to be fourteen pounds per square inch—water boils at a temperature of two hundred and twelve degrees Fahrenheit. If the pressure is lower, the boiling point will also be lower. Conversely, as the pressure increases, the boiling point of water increases. This is the key factor exploited by the designers of a pressurized-water reactor, who want to run high-powered reactors at high temperatures. So long as they make sure to maintain a high pressure of twenty-two hundred psi, water at six hundred degrees will not boil. There are, of course, many other possible combinations of temperature and pressure. Standard "steam tables" have been published to provide scientists and engineers with a handy reference that shows, among other things, what "phase"—liquid or steam—water happens to be in for every temperature and pressure combination. The tables themselves, like the tax tables that accompany income-tax forms, are very detailed, and proper instruction is needed to interpret them correctly.

The training given to the TMI operators did not emphasize the importance of monitoring reactor-cooling-water temperature and pressure during emergencies, or the techniques needed to recognize boiling conditions inside the reactor, such as the use of standard steam tables. The pressure inside the reactor during normal operation is always high enough to prevent boiling, so the operators generally don't need to give the possibility of boiling a second thought. Since their emergency training neglected this issue, they carried over their customary disregard of it into their emergency response on March 28, 1979. Although they knew that the pressure was abnormally low—it fell to one thousand psi within minutes after the accident began, and declined to as low as seven hundred psi within the first few hours—they completely failed to see the dangerous implications of this fact: that the low pressure would permit the hot water inside the reactor to boil away. Whereas a doctor would know immediately from a patient's very low pulse and blood pressure that a critical emergency existed, the TMI-2 operators had no such reflex reaction to the low-pressure readings they received during the accident. They found them merely puzzling.

If the operators had consulted a steam table, they would have been alerted to the boiling in the core, and they would also have learned something else of decisive importance. As the TMI-2 accident evolved, the temperature of the steam coming out of the reactor was far higher than the normal boiling point for the pressure that existed. The temperature was so high, in fact, that it indicated the presence of superheated steam inside the reactor, that is, steam at a temperature above the normal boiling point. There is only one way that superheated steam could be created inside the reactor: the core had to be uncovered. As the water level in the reactor fell, the tops of the twelve-foot-long fuel rods would become exposed. Then, as water lower down in the core boiled, the resulting steam, rising over the bare fuel rods, would be heated still further: superheated. The high temperature readings in the pipes coming out of the reactor, if compared with the normal boiling points in steam tables, would immediately show the superheated condition—a phenomenon, however, that even the plant's senior engineers, who reviewed conditions in the plant in the first few hours after the accident began, failed to recognize. The pervasive ignorance of basic engineering principles on the part of the TMI-2 staff was noted in a study of operator qualifications done for the presidential commission that investigated the accident. According to this study, this lack of basic technical competence existed at all levels of the TMI-2 staff, from the operators up to the plant engineers. "The 'last line of defense,' the engineer, is perhaps less qualified than the persons who turn to him for directions," the study concluded.

* * *

As with all other aspects of nuclear-plant safety, the ultimate responsibility for overseeing the selection and training of reactor operators belongs to the N.R.C. The Atomic Energy Act of 1954, which established the legal framework for the commercial nuclear-power program, specifically requires that the operators at every nuclear plant must be licensed by the federal government and states that federal regulatory authorities shall "prescribe uniform conditions for licensing" and "determine the qualifications of such individuals." When Congress was drafting this portion of the law, the A.E.C. actually tried to get the Joint Committee on Atomic Energy to delete it and to eliminate all requirements for federal licensing of reactor operators. The A.E.C. chairman at the time, Lewis L. Strauss, an investment banker who served as a member of the A.E.C. from 1946 to 1950 and was later appointed chairman by President Eisenhower in 1954, told Congress that he would prefer to leave operator training up to the utility companies. Like other members of the commission at the time, Strauss was an opponent of "bureaucratic red tape" that might "strangle" the infant nuclear industry, and he felt that the utility companies themselves were well qualified to oversee the training of their own staff. Congress was adamant and assigned the task to the A.E.C. anyway. As is often the case, however, bureaucratic intransigence is not so easily overcome. The A.E.C. accepted responsibility for operator licensing, and then did essentially what Strauss had wanted to do in the first place: they established a small staff to issue the licenses, but delegated responsibility for candidate selection and training to the individual utility companies without the benefit of governing regulations from the A.E.C. This situation did not change even as the nuclear industry matured and expanded.

A study on this subject by the presidential commission observed:

> Perhaps the most significant feature of the regulations is that there is no regulation regarding operator selection and training. The N.R.C. has no minimum eligibility standards for the qualification of operators. Rather, the N.R.C. endorses a standard established by the American Nuclear Society [an organization composed chiefly of engineers who work in the nuclear industry] pertaining to the selection and training of nuclear power plant personnel. This standard ... includes *recommendations* to the utility concerning selection, training, and qualifications. Reactor operator candidates do not have to meet any *requirements* concerning minimum education, experience, reliability, criminal record, or stress fitness.

Besides "a high school diploma or equivalent"—the term "equivalent" is not defined in the industry standard—the American Nuclear Society recommended that the candidate have a minimum of "two years of power plant experience or its equivalent provided that a minimum of one year is at a nuclear power plant." The standard does not define the term "experience," so no matter how irrelevant the candidate's prior work at a power plant might be to the duties of reactor operator, it can be counted as "power plant experience." Prospective control-room operators have generally worked as unlicensed auxiliary operators prior to their selection as control-room operator trainees. At TMI-2, as a result of negotiations with the employees' union, Metropolitan Edison had a policy of selecting the *most senior* auxiliary operators as operator trainees without determining which individuals happened to be the *most qualified.* In addition, the N.R.C. allows candidates who served in the Navy's nuclear program to count that experience toward the minimum expected of commercial reactor operators, however different the small naval reactors may be from a large commercial reactor. Moreover, the N.R.C. does so regardless of the candidate's type of duty in the Navy and without reviewing records of the candidate's performance. A medical examination of each applicant is required, to attest to the candidate's general good health, and the candidate as part of his medical history is asked to disclose any mental illness or nervous disorders he may have had. The N.R.C. does not verify this information nor does it require any screening to assess the candidate's ability to cope with the stress of emergencies or to determine whether the candidate may have any undisclosed psychiatric disorders. It is believed that a fatal accident at an A.E.C. test reactor in 1961 was caused by an unstable operator who intentionally destroyed the reactor. He killed himself and two other operators. This incident led Dr. Hanauer to recommend mandatory psychiatric screening of future reactor operators, but this recommendation has not been implemented.

* * *

Besides neglecting to specify detailed operator selection criteria, the N.R.C. also fails to set any meaningful rules to determine how operator candidates, once selected, will be trained. Consistent with the agency's policy of "industry self-regulation," the responsibility for operator training is delegated to the individual utility companies, which are presumed to be capable of developing and administering suitable programs. The N.R.C. does check to see that some form of training program is carried out, but since the agency's inspectors lack specific standards against which to judge the adequacy of the companies' training programs, the reviews are a *pro forma* exercise. The N.R.C. inspectors do not customarily bother to audit the training lectures that the companies give prospective operators, for example, but limit their "inspection" of the training program to a check of the records documenting the scheduling and other administrative aspects of training sessions; there is no substantive review of the content of the training program.

The broad discretion the N.R.C. allows the companies is further indicated by the agency's failure to specify any required qualifications for the utility-company instructors who give reactor operators their basic training. (In contrast, the Federal Aviation Administration, which licenses pilots, requires that the instructors who train the pilots pass a special F.A.A. flight instructor's licensing test.) Since instructor qualification is left up to each utility company, the instructors are generally no more qualified in terms of educational and professional background than the trainees. At TMI-2, for example, the members of the operator-training department, like the operators themselves, are high-school graduates who lack degrees in engineering, nuclear sciences, or any of the other specialized fields relating to nuclear-plant safety.

The respective corporate and bureaucratic "organization charts" of Metropolitan Edison and the N.R.C. provide a practical indication of the importance that each organization attached to operator training. At Metropolitan Edison, the department responsible for training operators was a disconnected unit, an established but seemingly forgotten branch of the company. In the conventional organization chart, the relationship of one company office to another is generally displayed by noting how one office or division "reports to" another. However, as discovered by an investigator working for the presidential commission, "At the time of the TMI-2 accident, the Training Department [of Metropolitan Edison] did not, in fact, report to anyone on the Island." Instead of being integrated with other plant activities under the control of

TMI-2 managers, the TMI training department, through an apparent fluke, reported to the "manager of quality assurance" at Metropolitan Edison's home office in Reading, Pennsylvania. That company officer was principally responsible for quality control during plant construction and for plant security, and he has told accident investigators that he visited Three Mile Island only about once a month and was unable to devote much time to operator training. After the TMI-2 accident—in May 1979—responsibility for operator training was reassigned to the superintendent of TMI-1 in order to put this task under the general supervision of on-site management. Yet, when interviewed subsequently by accident investigators, the presidential commission staff reported, "neither the station manager nor the superintendent of TMI-2 were aware of this."

The N.R.C. organization chart, similarly, shows the office concerned with operator qualification, the Operator Licensing Branch, to be a small, neglected division of the agency. It is located, to describe its situation in bureaucratic terms, in the "sixth echelon" below the commission, according to the presidential commission. As Commissioner Victor Gilinsky describes it, the Operator Licensing Branch is "in the basement" underneath the agency's Division of Nuclear Reactor Regulation. The office, headed by Paul Collins, employs eight full-time examiners who administer tests to control-room-operator candidates; there are another twenty-two "consultants" hired by the office to serve as examiners on a part-time basis. The Operator Licensing Branch is principally concerned with administering written tests to prospective operators and has little involvement with operator selection and training. Few of the examiners have any experience as power-reactor operators, and their main job is to administer and grade tests—not a particularly demanding task since the questions on each exam are generally chosen from a "bank" of some four hundred questions.

* * *

The reactor-operator-training programs established by the utility companies cover a large large number of subjects in a superficial manner. The standard utility-run program for operator training, according to a recent N.R.C. study, consists of: (1) a twelve-week introductory course on the principles of reactor operation, radiation and its effects, and "the necessity of operating a reactor in a responsible manner" (included also in this phase of the training is a "one-week laboratory course at a research reactor"); (2) a six-week lecture series on the generic operating characteristics of the facility for which the trainee seeks an operator's license; (3) a four-and-a-half-month program in which the trainee "operates the controls of a nuclear-power-plant simulator during normal, abnormal and emergency conditions" and observes "day-to-day plant operations" in an operating nuclear-plant control room; and (4) a one-year program at the facility the applicant will operate "to attend classes on the design features of the facility, write operating procedures, perform construction checkouts and run preoperational tests of equipment."

The emphasis in the standard training program is on procedures that the operators should follow rather than on principles they should understand. The procedures themselves are the standard techniques needed to operate the plant and to cope with possible plant malfunctions. Since all likely malfunctions are presumed to have been studied in advance by plant designers, and the results of this analysis incorporated into written procedures, the operator's basic training for his emergency duties is to memorize the preestablished procedures that he is supposed to follow. Basic knowledge needed to understand accident phenomena—information that on some unforeseen day might make the difference in the operator's ability to control a major emergency—is not emphasized. The goal of the training program is more limited: it is to train the operators to pass the N.R.C. operator-licensing examinations.

The reactor-operator-licensing examination given by the N.R.C. consists of an eight-hour written test followed by a subsequent oral examination administered in the plant's control room. The N.R.C. examinations generally call on the candidates to recite factual details that they have memorized about the plant they want to operate and to demonstrate that they have memorized plant-operating procedures. (An operator's license from the N.R.C. is not a general-purpose certificate permitting the licensee to operate any nuclear-power reactor; since there is little control-room standardization, an N.R.C. control-room operator's license is valid only for a specific nuclear plant. Even TMI-1 operators, for example, have to be relicensed if they wish to operate TMI-2.) Apart from asking candidates about the procedures to be used in handling certain simple accidents, there is no effort to probe the prospective operators' ability to diagnose more complex emergencies. Each examination does include five questions on the "principles of reactor operation," but these generally appear to challenge the applicants' ability to remember the definitions of concepts unlikely to play a role in any practical emergency situation they would ever face. There are, to be sure, "What if . . .?" questions on the examination, which ask applicants to describe the consequences of various equipment malfunctions. The malfunctions specified, however, are typically straightforward, single failures of nonsafety equipment (What happens if one of the main coolant pumps stops?) rather than more complicated cases (What do you do if certain key safety devices malfunction during an accident?). Seriously abnormal difficulties, such as a common-mode failure that completely disables the emergency feedwater system, are not presented to the applicants for analysis, nor is there an effort to review applicants' knowledge of basic physics and engineering principles that might be helpful in diagnosing major problems of this kind. A recent review of the standard N.R.C. operator-licensing-examination questions noted that several questions began with "what" or "how many," and additional questions asked the trainees to make "lists" of plant systems or to draw diagrams of plant equipment. Few questions—only two out of twenty-eight in the sample—asked the operators to explain "why" some phenomenon would occur during plant operation or during an accident. According to a report by the presidential commission, the examinations promote "a shallow level of operator knowledge" because they emphasize "memorizing numbers," how to manipulate controls, and "an elementary rather than a comprehensive" understanding of reactor technology.

Another study, by the Essex Corporation, reviewed the N.R.C. examination given to the TMI-2 operators to determine how closely the questions on the examinations related to the specific skills that the operators needed in responding to the accident. In order to execute six required procedures applicable to the accident, the study determined, the operators needed some fifty-three specific skills. These involved both the skills necessary to diagnose the accident and the knowledge necessary to carry out appropriate remedial actions. The typical examinations given to the TMI-2 operators, however, included only one or two questions relating to the fifty-three necessary skills. A satisfactory grade on the N.R.C.'s examination provided no indication whatever of an operator's ability to cope with the accident. The Essex Corporation study drew this conclusion from its review of the examinations:

> The approach to test construction appears to be too informal as is the approach to the entire training program. Tests should reflect job requirements and should be based on specific training objectives. They are not. Tests should be performance based and criterion referenced. They are not. Tests should comprise methods of presenting feedback to operators as to their performance strengths and weaknesses. They do not. Tests should measure both operator knowledge and skills, including the capability to diagnose a transient and identify causal factors, the capability to control plant systems, the capability of following procedures, the capability to anticipate the response of slowly reacting systems, and the capability to understand what is going on in the plant. They do not.

The N.R.C. examination for a prospective reactor operator is in seven parts. A passing score is an overall grade of seventy percent. It is possible, however, to do poorly, or even to fail individual parts of the test, yet to raise one's average to a passing grade by doing better in other sections. The N.R.C. does not mention in any of its regulations how it expects an operator who receives a passing grade as low as seventy percent to learn the thirty percent of the information that he missed on his examination, nor does the N.R.C. customarily give the utility company's training department copies of the graded examination so they might provide needed remedial training. The N.R.C.'s attitude toward the mistakes that prospective operators make on their written examinations is much different from that of the Federal Aviation Administration, which examines and licenses prospective airplane pilots. The F.A.A. has a standard procedure relating to wrong answers: the examination report sent back to the applicant gives, in addition to the overall grade, the specific technical subject areas in which the applicant answered questions incorrectly. Prior to receiving a pilot's license, the applicant must receive additional training in these subjects from a flight instructor, who must certify that the applicant has mastered the subject matter in question. "We don't do this, we don't have any feedback to the operators on wrong answers," the N.R.C.'s chief examiner, Paul Collins, acknowledges.

The deficiencies in the training of reactor operators would be less worrisome if the operators were working under the direct supervision of individuals more qualified in terms of educational background and training than they themselves are. The senior operators are required to pass a twelve-part examination: parts one through seven, the same examination given to operator's license applicants, and an additional five-part test. Some additional questions relate to "reactor theory," but according to the guidebook that the N.R.C. gives to applicants, these questions are "not advanced to the level of a nuclear physicist or engineer." Basic thermodynamics, for example, which is omitted in the operator's examination, is similarly omitted in the senior operator's examination. The passing grade on the senior operator's licensing examination is also seventy percent, and like the applicants for an operator's license, senior operator applicants can fail parts of the test and still receive an overall passing grade. It is even possible for those applying for a senior operator's license to flunk any of the first seven parts of the exam—the portion also given to operator's license applicants—and still receive a senior operator's license. One of the senior operators at TMI-2 did just this: he received an overall grade of eighty-four percent on his senior operator examination, but flunked the portion of his operator license examination that governed "general operating characteristics" of the TMI-2 plant; his grade in that category was sixty-two percent.

* * *

Despite their limited scope, the examinations given to the operators and senior operators at TMI-2 did include several sets of questions designed to see whether the applicants had "at their fingertips" a multitude of elementary facts and simple details about the plant. Copies of the actual examinations taken by TMI-2 personnel, as graded by the N.R.C. examiners, have been made public as a result of a Freedom of Information Act request made by Robert Schakne of CBS News. The names of the applicants were withheld to protect their privacy. The candidates at TMI-2, according to the N.R.C., scored above the national average. Still, a review of the examinations shows a notable spottiness in the operators' knowledge of the plant they were being licensed to operate. The tests also show a considerable unevenness in the level of knowledge among the individual operators at the plant.

Some operators, for example, seemed to be intimately familiar with the specific plant systems about whose design or performance they were questioned. One applicant for a senior reactor operator's license, however, could not draw a simple diagram showing how one of the main electrical-power-distribution systems at TMI-2 received its power supply. His grade for this question was zero. Asked a related question, how this power distribution system would work during a common emergency condition, his grade was thirteen percent. Asked how the pressurizer-water-level indicator would behave if a reactor coolant pump failed when the plant was at full power, this applicant once again received a grade of zero. Nevertheless, he received a senior operator's license. Another applicant for a senior operator's license flunked the portion of his examination on fuel handling and core parameters and received marginal grades in reactor theory, radioactive materials handling, and the specific operating characteristics of TMI-2, but on administrative procedures he received a grade of ninety percent. His overall grade was seventy-six percent. The N.R.C. examiner who graded this written examination and gave the applicant an oral examination approved his application for a senior operator's license and added a comment on the examination report: "Good man for shift supervisor."

In several similar instances, the examiners noted the applicant's significant weaknesses but nevertheless issued an operator's license or a senior operator's license. One applicant was given an operator's license even though his examiner noted, "Marginal in normal and emergency procedure also in auxiliary room equipment." In another case, in which the applicant flunked the examination section on the general operating characteristics of TMI-2 and received a marginal grade in standard and emergency operating procedures, the examiner signed the examination report recommending an operator's license and added the comment, "Impressive person, appears very knowledgeable, conscientious and thorough." Another applicant drew this examination report comment from the N.R.C. examiner: "Applicant on site for only 6 months. Not familiar with control room. Progress very slow and answers had to be led out of him. Not a bad job outside of control room. Unable to apply theory to practical situations." This applicant, too, received a reactor operator's license.

Of particular interest in the Three Mile Island operator examinations are the questions that relate to the specific problems that arose at the plant during the accident. Several TMI-2 operators, for example, could not provide correct answers to the question of what they would be required to do during a reactor trip (such as occurred on March 28, 1979). Others were unfamiliar with the reactor coolant-drain tank and the instruments that describe its condition (instruments that might have helped diagnose the stuck-open relief valve). One applicant for a senior operator's license was questioned about one of the very conditions that happened during the accident:

> It has been established that a small RCS [Reactor Cooling System] leak exists into the reactor building. Assuming that the leaked water collects in the building sump, describe, in sequence, what happens to the water from the sump to its ultimate disposal.

The applicant could only describe a small part of "what happens" to water after it gets into the sump. His grade on this question was thirty-three percent. This was not his only weak spot; he received a grade of seventy-two percent for his knowledge of the general operating characteristics of TMI-2, a grade of seventy-eight percent in standard and emergency procedures, a grade of seventy-one percent in reactor theory, and he flunked the examination section on "radioactive materials handling, disposal, and hazards" with a grade of sixty-four percent. This marginal applicant was nevertheless awarded a senior operator's license. According to the N.R.C. regulations, TMI-2 is legally authorized to operate at full power without any more qualified person in charge than this senior operator.

* * *

Operating a nuclear reactor requires practical skills that cannot be required in a training program restricted to classroom lectures and cannot be measured by written examinations. The trainees must get actual "hands-on" experience at the controls in order to learn basic operating techniques and to gain familiarity with the response of plant systems to operator actions. However, since a nuclear plant in normal operation runs under automatic control, even on-the-job training in the control room will not amount to much: the trainees, like the operators, just get to sit and watch. Observing the control room during normal operation, moreover, the trainees will be unlikely to see the type of infrequent accident conditions that they must be prepared to handle. These factors underscore the importance of the training devices, known as control-room simulators, that are now a standard part of the reactor operator's training program. The simulator—a realistic mockup of an actual nuclear-plant control room—uses a special computer to make the instruments and signal lights on its control panels behave like those on an operating reactor. (The setup is similar to that in the cockpit simulators that have been used for years to train airline pilots.) The computer is programmed to simulate actual control-room performance during a wide range of circumstances and to produce a realistic response when the operator trainee manipulates the controls. For example, when the trainee uses the controls required to start up the reactor, the computer would make the instruments on the simulator's panels show the corresponding increase in the reactor's power level. If the trainee makes a mistake, the actual consequences of the error will be reflected in the responses of the instruments. The computer can also be programmed by the instructors to simulate various types of emergencies so that the trainees can see how the control-room instruments would behave in such a situation and how their own actions at the controls would affect the outcome of the simulated crisis.

Because simulator training is potentially one of the most useful parts of the operator's training program, the N.R.C. encourages the use of these devices. The agency sets no specific requirements, though, on just what the operators ought to do in simulator-training programs, nor does it require that any minimum amount of time during their training program be spent on simulators. The TMI-2 operators, like most reactor operators in the United States, spent very little of their training time on the simulator. About eighty-five percent of their training time was spent in the classroom, ten percent was spent in on-the-job training, and only about five percent was devoted to the use of simulators, according to a recent study done by the Essex Corporation. The Metropolitan Edison Training Department did not have its own control-room simulator; this is an expensive item, and few of the utility companies that operate nuclear plants have one of their own. Instead, simulator training is usually provided as a service by the manufacturer of the plant's reactor. The TMI-2 operators, accordingly, were trained on a control-room simulator at Babcock & Wilcox's Virginia headquarters. This was actually a mockup of the control room at the Rancho Seco nuclear plant in Sacramento, California, but it contained all the important controls and instruments installed at TMI-2 (though not necessarily in the same locations). The simulator was capable of showing the operators what it would be like in the control room during startup, normal reactor operation, routine shutdowns, and various emergency conditions. The simulator also had a "freeze" control so that the instructors could stop the action, discuss with the trainees what was happening, and explain to them what they ought to be doing about it.

The simulator training given to the TMI-2 operators was conducted by the Training Services Section at Babcock & Wilcox, a part of the company's Customer Services Department. None of the instructors in the Training Services Section had an engineering degree, but all were former senior reactor operators, although there was no requirement that they periodically requalify. (The F.A.A., in contrast, requires licensed flight instructors to be reexamined periodically in order to demonstrate that they have kept current on developments in the field.) The Training Services Section did not offer a systematic overall training program for the personnel who would operate Babcock & Wilcox reactors. Instead, it offered a set of short package courses. The customer utility companies could send their personnel to whatever courses it wanted them to attend. Some of the courses offered simulator training for new operators, and others offered refresher courses for already licensed operators. In addition, a variety of specific lecture courses and seminars, as well as more general "orientation" courses, were available for plant technical personnel. Unfortunately, the training programs offered by Babcock & Wilcox did not reflect the company's expertise with the specialized nuclear equipment it designs and sells. The Training Services Section at Babcock & Wilcox—like the training department at Metropolitan Edison and the Operator Licensing Branch at the N.R.C.—was a satellite within the company whose usual orbit intercepted none of the other technical divisions in the organization. There was no coordination between the engineering staff and the training staff to develop an appropriate training program. Company safety experts, who spend their careers assessing the phenomena that might take place during hypothetical accidents, had no involvement in setting up a training program to teach operators to recognize those phenomena. Babcock & Wilcox engineers have admitted that they themselves have almost no practical experience with the operation of the machinery they design and no substantive involvement in operator-training programs. One senior Babcock & Wilcox engineer told the presidential commission that he was unsure if there was an operator-training section in the company, and the vice president in charge of the company's nuclear-power division admitted that little management attention was devoted to the training program. "Nobody in B & W management above the Training Services Section has given attention to course administration," according to a study done for the presidential commission. There is only one evident sign of management interest in the Training Services Section at Babcock & Wilcox: the fees for the courses appear to have been set high enough to cover the company's direct costs.

The simulator-training program given to the TMI-2 operators by Babcock & Wilcox was of little assistance to them during the accident. For example, it never covered the case of a stuck-open relief valve (and how to recognize and correct the problem). None of the simulator exercises covered a situation in which the reactor's cooling water reached saturation conditions and started to turn to steam—the simulator, as then programmed, was incapable of simulating saturation. Nor was the simulator programmed to reproduce the strange behavior of the pressurizer-water-level instrument that materialized during the accident: the high reading that the instrument displayed even though the reactor actually had a dangerously low quantity of required cooling water. There was no emphasis in the program on the necessity of keeping the core covered with water, nor were the symptoms of an uncovered core, such as superheated steam, taught to the operators.

The Babcock & Wilcox simulator was a sophisticated and flexible training tool, and its computer could have been programmed to reproduce the symptoms of a very wide range of complicated accidents. As programmed by the Babcock & Wilcox Training Services Section, however, its repertoire was limited to a set of eighty relatively simple and straightforward malfunctions. The operators, as Ed Frederick recalls, were shown cases in which one of the three emergency feedwater pumps malfunctioned. This "accident," however, created no dire emergency because the other two pumps by themselves were fully capable of supplying an adequate amount of emergency feedwater. A case in which the entire emergency feedwater system is disabled simultaneously, as happened during the accident, was never programmed into the simulator. Such a failure, Frederick noted, "was against the rules." In general, the simulator exposed the trainees only to simplified accidents in which the emergency equipment started up automatically and worked successfully. They were trained how to "monitor and verify" the performance of the emergency systems. Although intended, in principle, to put the operators through full-dress rehearsals for possible accidents—the simulator was even programmed to include the background noise that the operators might hear—the computer was actually set up to reproduce only accident conditions in which the safety systems work satisfactorily. The "basic assumption" behind their simulator training, Ed Frederick has observed, was that the type of accident that struck TMI-2 "probably couldn't happen." The operators had received specific training on how to deal with sudden plant shutdowns, he said, but he added:

> Every time we approached this type of accident, the outcome was assumed. In other words, the safety systems all activated, the emergency feed worked ... and although they would throw in a few instrument errors once in a while, the basic philosophy was: everything is going to work and it will be okay.

The Babcock & Wilcox simulator-training program repeated the same set of simple emergency exercises year after year. The exercises were not systematically updated to reflect the accidents that were actually taking place at Babcock & Wilcox plants or to reflect new knowledge in the reactor-safety field. Thus, after the Davis-Besse accident of September 24, 1977—an event that attracted the attention of other Babcock & Wilcox departments—no revised lesson plan was developed for the operator trainees so that they would be shown this accident on the simulator. Babcock & Wilcox officials, including the manager of emergency-core-cooling analysis, Bert Dunn, wrote internal memorandums after the Davis-Besse accident on the need to give operators better instructions on handling similar types of loss-of-coolant accidents. Some of the memos were sent to the Customer Services Department, which oversees the Training Section, but the recommendations were never incorporated into revised operator-training programs. Even months after the TMI-2 accident, Babcock & Wilcox instructors, when questioned by the presidential commission, had still not heard of the memos on the Davis-Besse accident.

There are three different ways in which a simulator can be used as a training device, according to Essex Corporation analysts who reviewed the Babcock & Wilcox simulator training program. First, it can be used to demonstrate aspects of plant behavior to the trainees, who just stand there and watch. Second, it can be used to put the trainees, or the operators taking refresher courses, through practice sessions in which they learn how to diagnose and respond to particular types of emergencies. By repeating the drills, and learning from their mistakes, they can acquire basic skills and the confidence to use them. Finally, the simulator can be used to evaluate and measure operator performance, checking to see how well the trainees diagnose accidents, how reliably they carry out necessary emergency actions, and how quickly they are able to perform. The Essex Corporation concluded that necessary simulator training for the TMI-2 operators was "largely ignored," since only five percent of the training time was spent on the simulators. The Essex Corporation study explained:

> This conclusion is based on the fact that the shift supervisor on duty at TMI at the time of the accident had undergone re-qualification training on the B & W simulator in March 1979. During the 20 hours of simulator operation, a total of 19 different ... emergencies were simulated. Of these, 14 were only performed once and only one was performed as often as three times. This would indicate that the simulator is being used to illustrate for the operator what a selected emergency looks like in terms of display readings and plant reactions. It is not being used to allow the operator to acquire skill through practice in responding to faults and formulating hypotheses concerning what is happening in the plant.

Another study, done for the presidential commission, noted that as many as fifteen simulated emergencies per hour were being run on the Babcock & Wilcox simulator. These drills were being run so quickly that only the immediate action that takes place at the beginning of the accident was shown to the operators. Since the events were not carried through to "their logical conclusion," the trainees were not acquainted with the longer-term responsibilities they had. "For instance," the study noted, "a loss-of-coolant accident might not be carried past the point of [emergency-core-cooling system] actuation" during the simulator exercise. During the TMI-2 accident, there was an "actuation" of this basic emergency cooling system two minutes after the accident began. While the simulator drill instructor may have considered the "accident" over at that point and gone on to other exercises, it was not over at that point on March 28: the operators, two minutes after the system came on, shut it off—not having been properly trained to leave it on.

Babcock & Wilcox instructors used the simulator merely to "illustrate" what accidents look like from the control room. They made no use of the simulator to evaluate operator performance, and did not notify the client utility company when an operator from that company performed unsatisfactorily on the simulator. Nor did Metropolitan Edison audit the program to determine its adequacy or send company officials to observe and evaluate the performance of TMI-2 operators and operator trainees who were sent to use the simulator. The N.R.C. itself, though it recommends simulator training, does not require that prospective operators demonstrate their proficiency on the simulator in order to receive an operator's license. The N.R.C. awards licenses without requiring any practical demonstration of the applicants' ability to respond satisfactorily in an emergency situation. The N.R.C. does require, however, that licensed operators, in order to keep current, perform a minimum number of control manipulations every two years in order to maintain their proficiency, and it permits the operators to meet these requirements by manipulating the controls of a control-room simulator. The practice at Babcock & Wilcox was to give all simulator trainees who were present during such an exam credit for the manipulations whether they themselves operated the controls or not. The N.R.C. was unaware that this practice was in effect.

In an actual emergency, the operators on duty must function as a crew. This is especially so since the random assortment of instruments and controls around the room may require operators at different positions in the room to perform coordinated actions, one reading a distant instrument while another manipulates a control, for example. Trainees on the Babcock & Wilcox simulator, however, did not go through emergency drills as a crew. Shift supervisors, who would be required to manage the crew in an actual emergency at TMI-2, did not train with a crew to prepare themselves for carrying out this assignment.

* * *

Once licensed as a reactor operator or as a senior reactor operator by the N.R.C., an individual will have no further routine contact with the agency. The license must be renewed every two years, but license renewal is not predicated on passing any further N.R.C. examinations or special proficiency checks. The N.R.C. simply specifies that operators and senior operators should complete a "requalification program" administered by the utility company for which they work; license renewal then becomes automatic.

The core of the typical operator requalification program consists of lecture series conducted by the utility company training departments. The lectures given to TMI-2 operators were superficial and contributed little to their general basic knowledge. One potentially useful part of the plant's lecture program was a review of accidents that had occurred at other plants, including those designed by Babcock & Wilcox. The purpose of lectures on this topic was to help TMI-2 operators become familiar with potential problems and how to handle them, and to avoid any mistaken responses that might have been made by the operators involved in these prior incidents. According to the presidential commission, the TMI-2 operators spent only about one hour per year in useful lectures of this type. The one lecture they attended, moreover, "was devoted to discussion of relatively minor material problems such as with reactor coolant pump snubbers, reactor building doors and diesel engines," according to the commission. Accidents in which operator error played a role were not discussed, and no mention was made, for example, of the significant mishap that had occurred at Davis-Besse on September 24, 1977, the event that so closely prefigured the accident at TMI-2. (The brief summary report of this incident prepared by Toledo Edison and sent to the N.R.C., a copy of which Metropolitan Edison also received, did not mention the significant aspects of the event, such as the operator's mistaken decision to shut off the emergency cooling system, and the Metropolitan Edison officials who prepared the one-hour lecture on the recent "operating histories" of other plants did not include a reference to the Davis-Besse accident because they had no way of evaluating its significance.) Even when lectures of some usefulness were offered to the TMI-2 staff, the operators themselves were often too busy to attend. At TMI-2, attendance at the lectures was less than fifty percent, on the average, and senior officers, who had more administrative duties, were even less conscientious than operators about attending the sessions. According to an October 1979 report by the presidential commission staff, the superintendents and the supervisors of operations at both TMI-1 and TMI-2, all four of whom are licensed operators, had attended none of the training programs at the plant during the previous year. Personnel who missed the lectures were given "care packages" by the TMI training department so they could cover the material from the lecture programs on a "self-study" basis.

The N.R.C. does not do a detailed review of the annual evaluation examinations of plant operators and senior operators. The agency's normal procedure is to have an N.R.C. inspector "spotcheck" a few of the exams. N.R.C. officials say that the annual examinations for about six TMI-2 operators had been audited by an N.R.C. inspector, although the agency's records contain no documentation describing the results of this review. In practice, the N.R.C. leaves the evaluation of the examinations up to the utility companies and relies on their determinations as to whether the individual operators and senior operators have maintained their proficiency. The tests administered by Metropolitan Edison do not appear to have been a great challenge to the TMI-2 staff; all but two licensed operators at the plant passed the annual examination given to them in February 1979. One person who flunked the examination, however, was the supervisor of operations at TMI-2, one of the most senior management officials at the plant. By the time of the accident, he had not yet participated in any remedial requalification work.

The N.R.C. licenses reactor operators who directly manipulate the reactor controls, but does not require the licensing of the engineers and managers who give direction to the operators and oversee the operations of a nuclear plant. These higher-level technical and managerial personnel will inevitably become involved in responding to a serious nuclear-plant emergency, but there is no check to make sure that they have the requisite knowledge and experience to manage the plant in such circumstances. It is "recommended" by the American Nuclear Society that the plant manager "shall have acquired the experience and training normally required by the N.R.C. for a senior reactor operator license," even though he may not actually hold a senior operator's license, and that the operations manager shall hold a senior operator's license.

The shift supervisor is the first member of plant management to become involved in handling a crisis. This is not, however, a role for which officials who hold that position are carefully prepared. Like William Zewe, who was on the overnight shift at TMI-2 on March 28, shift supervisors are typically high-school graduates who have not been given any special training for handling major emergencies. In the naval nuclear program, the reactor operators are supervised by an "engineering officer of the watch" who is specifically trained for the role of providing overall guidance in diagnosing and responding to serious reactor problems. In the civilian program, the shift supervisor assumes this responsibility without the benefit of an engineering degree or special training. In fact, the basic job of the shift supervisor could hardly be less suited to a role as emergency decision maker. The shift supervisors at TMI-2 estimate that somewhere between fifty and eighty percent of their time is spent on routine paperwork. As one N.R.C. task force on the "lessons learned" from the TMI-2 accident observes:

> The day-to-day routine of many shift supervisors has become increasingly devoted to administrative details. Instead of providing direct, command oversight of operations and performing management review of ongoing operations in the plant that are related to safety, they find some of their time devoted to lesser chores. Their activities can range from the scheduling of overtime and meal money to review of radiation work permits, maintenance requests, and surveillance procedures.
>
> Many shift supervisors have attained their position after having served for a number of years as a control-room operator; that is, a direct manipulator of plant controls. Although the training received by senior operators [the rating held by shift supervisors] is directed at increasing their technical understanding and knowledge of administrative controls of operations, no real emphasis is placed on their management or command role.

Zewe's performance during the accident bears out these observations. No more able than the two operators to analyze the technical troubles affecting the reactor, he completely missed the significance of the low-pressure symptoms exhibited by the reactor, and failed to provide overall management of the operators' response to the accident. Zewe was highly respected as a shift supervisor by the staff of the plant, who thought him able and hard-working, and he had received the highest scores of all the TMI-2 senior operators on the N.R.C. licensing examinations. His high grades but poor performance are testimony to the basic inadequacy of the training that he received.

Zewe was the most senior official on duty when the accident began. Above the shift supervisor, there are managers of many levels at a nuclear plant. Most were home asleep when the accident occurred but were called in over the next few hours. As plant officials ascend the management ladder at TMI-2, there is no necessary increase in their technical competence or knowledge of plant systems. To some extent, the exact opposite appears to be the case when one reviews the performance of TMI-2 plant management during the crisis. For example, fifty minutes after the accident began, George Kunder, the superintendent of technical support for TMI-2, arrived in the Unit 2 control room to help Zewe. Kunder, who was the "duty officer on call," was called to the plant not because of a perceived crisis, but to make routine postshutdown reviews of plant conditions. He had an engineering degree, fifteen years of experience in the nuclear program, and had held a senior operator's license for TMI-1. TMI-1 differs in many ways from TMI-2, and Kunder was still in training for a license to operate the new unit. He was unfamiliar with many of the details of the TMI-2 plant and felt uncomfortable about actually operating it. As a result, he stayed on the sidelines and was of no practical help in organizing an effective response to the accident. Despite a degree in engineering, he, too, failed to recognize the low pressure in the reactor as a warning sign that the core might be inadequately cooled, and he was of no assistance in diagnosing the fundamental cause of the trouble: the open relief valve.

Kunder's boss, Joseph Logan, the TMI-2 superintendent, arrived fifty-five minutes after Kunder. Logan, who was one rung higher on the TMI-2 organization chart, also had an engineering degree and twenty years of experience in the Navy's nuclear-power program (he was a retired captain and former commander of a squadron of nuclear submarines). Logan had received a senior operator's license for TMI-2 only about four months prior to the accident, and he lacked intimate knowledge of the plant. He also had no special training for managing nuclear-plant emergencies. Like Kunder, he was unable to play any substantial role in responding to the event or in diagnosing its causes.

Finally, at the top of the TMI-2 management chain of command was station manager Gary Miller. He had been notified of a routine, unscheduled plant shutdown just after it occurred and called back an hour and fifteen minutes later—around 5:15 a.m.—just to check in with plant officials before going out of town. He was then advised of the principal instrument readings noted by the operators, including the very low pressure in the reactor. Although he concluded that the readings were highly abnormal, he was unable to make a diagnosis of the actual problems facing the plant. He, too, was an engineer, but the engineering degree that he received from the Merchant Marine Academy in 1963 and his eight years as a testing engineer and as a construction manager for the Newport News Shipbuilding and Drydock Company before he came to Three Mile Island hardly provided pertinent background for diagnosing a complicated nuclear-reactor accident. Miller did at one time hold a senior operator's license for TMI-1, but he had very little actual operating experience and had never been licensed for Unit 2. Feeling "uneasy" about plant conditions after his telephone call at 5:15 a.m., Miller arranged to have a conference call set up at 6 a.m. so he could discuss the situation with John Herbein, a vice president of Metropolitan Edison in charge of the company's power-generation division, and Leland Rogers, the Babcock & Wilcox site representative at Three Mile Island. Kunder, who was at the plant, also participated in the conference call, which lasted for about forty minutes. Herbein was an Annapolis graduate and another veteran of the Navy's nuclear program; he was reached in Philadelphia, where he had gone for Naval Reserve duty. Rogers, who was at home near the plant, was a high-school graduate who had served as an enlisted man in the Navy's nuclear program but had never been trained to operate Babcock & Wilcox plants; he was more familiar with plant construction problems than with plant operation. The 6 a.m. conference call among these senior officials resolved nothing: they discussed plant conditions but were unable to analyze them in any conclusive way. They did note the very low pressure in the reactor but, like the operators, failed to grasp the significance of this conspicuous clue. Unable to diagnose the problem, they developed no emergency strategy and were not even prompted by the far-reaching uncertainties and the unprecedented conditions at the plant to notify the N.R.C. or to call for additional outside help. "In their confusion," Henry Kendall, a physicist at the Massachusetts Institute of Technology, has commented, "they were like children lost in the woods."

As the senior officials were discussing the situation on the telephone, Brian Mehler, the shift supervisor for the incoming 7 a.m. shift at the plant, who arrived about an hour earlier than usual at 6:18 a.m., suspected that the relief valve was open, and hence that the block valve should be closed thereby narrowly avoiding a meltdown. Kunder, however, who was in the shift supervisor's office next to the control room, was unaware of what Mehler and the other operators were doing, and the latter were unaware of the conference call. When Leland Rogers, sometime after 6:18, asked whether the block valve was closed, Kunder went out to check it and reported that it was closed to the others. No one asked when it had been closed, so the discussion proceeded without benefit of the knowledge that the block valve had just been closed and that the relief valve had been stuck open for more than two hours. The emergency cooling system was still shut off and no order came from the senior officials, who were unaware of the protracted loss of cooling water, to turn it back on. The senior officials continued to discuss the state of the plant until about 6:40 a.m., when they decided that Miller and Rogers should report to the plant as soon as possible.

Shortly after the call ended, a crisis still not formally acknowledged, high radiation levels in the plant automatically triggered radiation alarms in the control room. "I knew that we had a tremendous problem at that point," Zewe remembers, "so that is when I made the announcement and sounded the alarm that a site emergency had been declared in Unit 2." Radiation monitors in the plant's auxiliary building caused some of the alarms. Radioactive water from the reactor had overflowed tanks in the auxiliary building, flooded the lower parts of the building, and was backing up through the floor drains. Dale Laudermilch, one of the technicians working in the auxiliary building, has recalled that they were "sitting there trying to figure out where in the devil this water is coming from" and that he was thinking, "Man, oh, man, we are really going to be crapped up" because of all the radioactive material in the water. These reflections were interrupted, he said, when Mike Janouski, a radiation-chemistry technician, "came running down the hall and he just said, 'Get the hell out! Get your stuff and get out!' " The auxiliary building was evacuated, and fifty or sixty technicians, managers, engineers, and reactor operators crowded into the

TMI-2 control room, where the noise level and tension were very high. Still, no one, from the untrained technicians all the way up to the senior plant engineers, knew what was really happening to the plant, what the true peril was, or what any of them should be doing about it. This complete inability to deal with the situation reflected what one Essex Corporation study refers to as "the training disaster" that left TMI-2 personnel grossly unprepared for serious accidents.

All the plant employees could think to do at that point, three hours after the accident began, was to pick up the telephone and start calling outside authorities to warn them that TMI-2 had declared a site emergency. As they were beginning to make the appropriate calls, Gary Miller, the station manager, arrived in the TMI-2 control room. At 7:24 a.m., on the basis of high radiation levels in the reactor building, he declared a General Emergency. Among those notified were the state police, the state of Pennsylvania's Bureau of Radiological Health, the U.S. Department of Energy Radiological Assistance Plan Office, the Dauphin County Civil Defense Agency, and the American Nuclear Insurers Company. Plant employees also placed a call to the N.R.C.'s regional office, just outside Philadelphia, to inform the agency of what had happened. Unable to reach any N.R.C. officials there at that hour, they left a message with the answering service.

# 5: Crisis Management

Senior officials at the Nuclear Regulatory Commission reacted calmly, some even casually, to the telephone calls on Wednesday morning, March 28, 1979, informing them that the operators of the Three Mile Island Nuclear Power Station Unit 2 had declared a General Emergency. "I was going out of town that morning when word first came in," Harold Denton, the director of the Office of Nuclear Reactor Regulation, recalled during an interview a few months later. "I said, 'Ho, hum, I've got a more important meeting to go to.' " Denton decided not to go to the N.R.C. emergency center upon hearing the news but to send his deputy, Edson Case, instead. Dr. Joseph Hendrie, the N.R.C. chairman, was out of the office that Wednesday. He learned of the General Emergency then in progress when he phoned in from Washington Hospital Center, where he had taken his daughter to have wisdom teeth extracted. Dr. Hendrie decided not to go to the emergency center, or back to the office that day. In Hendrie's absence, Commissioner Victor Gilinsky became the acting chairman. Though a commissioner since 1975, Gilinsky was unfamiliar with the agency's emergency procedures; he did not go to the N.R.C. emergency center but remained in his own office in downtown Washington.

Even the N.R.C. officials who did go the Incident Response Center, which is located in an N.R.C. office on the East-West Highway in Bethesda, Maryland, just outside of Washington, were responding according to established procedures, or out of curiosity, rather than from any real sense of alarm. "I learned of the accident from my technical assistant at nine-twenty Wednesday morning," Commissioner Peter Bradford later said. "I decided to go out to the Response Center," he explained, "not at the time because I appreciated the seriousness of the accident but because I wanted to see how the Response Center functioned. I had not been there during an event or accident previously." Victor Stello, Jr., then head of the Division of Operating Reactors, was notified around 8 a.m., shortly after arriving at his office. He asked whether radioactive materials had been released into the environment. Told that this was believed to have occurred, he went to the Incident Response Center to join other members of the agency's "senior management team" who were expected to respond. "I don't know the thought I had in my mind at that moment," Stello said several weeks after the accident, "whether I would have called it an 'accident,' an 'incident,' or what—where it fell." Stello ordered other officials, such as Brian Grimes, an expert in accident-consequence analysis, to go with him to the center. "I had no particular reaction to the call," Grimes says, "I just went." One senior N.R.C. official—"speaking confidentially"—noted:

> I personally didn't expect such an accident to happen so early, but I have been somewhat of a pessimist and I did expect one sometime during my career. When I got to the Incident Response Center, I kept pinching myself to make sure it was the real thing. It was hard to believe it. We had exercises a couple of times a year when we're called to the Incident Response Center. Usually for exercises they will say, "This is an exercise." This time, they didn't say that.

Word of the accident quickly passed through the agency that morning. One N.R.C. staff engineer recalls hearing the news over the public-address system in the building in which he worked. "They announced that there had been an incident, that the emergency-core-cooling system was used, and that it worked," he remembered, "and everybody in the office applauded. People treated it all very lightly—like a joke or something—but then when things started to get worse, they stopped the announcements and all we knew was what we got from TV or newspaper reports."

The unprecedented news of a General Emergency at an operating nuclear-power plant produced a variety of emotional reactions among the senior officials of the N.R.C. Most of these men, engineers by training, had spent their professional careers in the nuclear-power program, and in the course of their duties for the A.E.C. and the N.R.C. over the years, some had written the general safety rules and regulations governing Three Mile Island; some had reviewed and approved the specific design features of the plant; and some had supervised its construction, the training of its operators, and the preoperational testing of its key components. The most senior N.R.C. officials had certified the plant as fit in all respects for safe operation and had issued its license to operate. Official blessings had been bestowed on the plant despite a broad array of doubts that had been expressed about aspects of its safety: specific technical criticisms from the N.R.C.'s own scientists and engineers, concerns expressed by some members of the general scientific community, and less well articulated but equally serious concerns from some of the citizens who lived in the vicinity. The N.R.C.'s Advisory Committee on Reactor Safeguards had warned in its 1969 "Report on Three Mile Island Nuclear Station Unit 2" of the susceptibility of "vital" plant-safety equipment to common-mode failures. Their own Instrumentation Task Force had warned about the lack of necessary control-room instrumentation in plants of this type to help reactor operators cope with potential accidents. Their own staff scientists had identified more than a hundred unresolved safety questions relating to pressurized-water reactors, the kind installed at TMI-2. All these concerns had ultimately been brushed aside, smoothed over by official rationalizations that the plant was still "safe enough," and the unit had been sanctioned by senior N.R.C. officials for full-power operation.

Once the severity of the accident became fully apparent, some N.R.C. officials felt guilty, others felt shamed and embarrassed, and at least one official, according to his colleagues, came close to having a nervous breakdown. No one in the agency was left unmoved by the event. These emotional reactions were somewhat delayed, however. On March 28, the day the accident began, the well-known human tendency of denial dominated the initial response of N.R.C. officials to the news: they tried very hard not to believe it.

"The conscience part didn't hit me until later, the feeling of responsibility, the awful feeling of responsibility," Dr. Roger Mattson explained during a long interview a few weeks after the accident. He had headed the N.R.C. division that had carried out the official safety review prior to the licensing of TMI-2. The N.R.C. officials, he said, tended to search for ways to explain away the data from the plant that morning—saying, for example, that the unwelcome reports reflected "instrument errors"—and refused to accept the evidence that a very damaging accident was in progress. "People still wanted to believe the best," Dr. Mattson said. "The N.R.C., for the first couple of days, wanted to believe something better was going on. The 'want-to-believe' mind-set evolved. Nobody would believe that what was actually happening could happen. I suspect this happened in the TMI-2 control room, too."

The records of the Incident Response Center bear out Dr. Mattson's observations. All the center's incoming and outgoing telephone conversations were tape-recorded on a twenty-channel recorder, the kind of device that police departments often use to record emergency calls. The thousands of small tape-cassettes from the center were transcribed by the N.R.C. at the request of accident investigators. Senior officials, according to the transcripts, shortly after arriving at the center and reviewing the early reports from the plant, quickly formulated an optimistic official assessment of the situation at TMI-2 and dismissed the contrary evidence. Their attitude toward the situation, moreover, remained fixed despite continuing technical reports that they received throughout the day, some of which suggested the possibility of severe damage to the reactor core.

One significant instrument reading that was relayed to the N.R.C. by telephone, early that morning, was the indicated radiation level in the containment building housing the reactor. Beginning at about 7:13 a.m., readings on a radiation monitor under the dome of the building began to increase rapidly. In a five-minute period, there was a spectacular two-hundredfold increase in the reported radiation level. This indicated that massive amounts of radioactive materials were leaking out of the reactor, which could only happen if the reactor core had been severely damaged. It was, in fact, the reading from this instrument that prompted station manager Gary Miller to declare the General Emergency at 7:24 a.m. N.R.C. officials were duly informed of this reading, which showed a radiation level of twenty thousand rads per hour (a lethal radiation dose is about four hundred and fifty rads). By about 9:30 a.m., besides this ostensibly disturbing information, N.R.C. officials in the Incident Response Center had learned of other signs of trouble, such as the presence of steam pockets inside the reactor's cooling system, which, as any nuclear-safety analyst knows, can prevent the core from being adequately cooled. Edson Case, one of the most senior N.R.C. officials analyzing the reports from the plant, noted in a 9:01 a.m. telephone conversation with Commissioner John Ahearne that part of the core had "apparently uncovered" at some point during the accident. Case said that the "real question" was whether plant officials had "regained control" over the reactor. Soon afterward, however, instead of continuing to explore the possibility that a major calamity had befallen the reactor, Case and other senior N.R.C. officials, like the personnel in the control room, began to discount the most worrisome evidence and attribute it to "instrument error" or other causes. This attitude was evident as early as 9:08 a.m., when Lee Gossick, the head of the Incident Response Center, told officials in the agency's downtown Washington office that the high radiation readings from the containment-dome monitor should be disregarded. Shortly after 10 a.m., three of the N.R.C. commissioners were briefed by the senior officials in the Incident Response Center. While passing along the available data to the commissioners, Edson Case reassuringly commented, "I think right now we have the situation under control, and we'll have to keep getting information to make sure that that continues."

The continuing reports from the plant provided no such confirmation that the situation was "under control," but this did not prompt Case to revise his earlier report to the commissioners. Quite the contrary. At his next briefing of the commission at around 11 a.m., Case announced that "the signs are encouraging" and "continue to be good." Then, between noon and 1 p.m., the Incident Response Center received data on the temperature and pressure of the water inside the reactor. If analyzed correctly, the superheated-steam conditions—evidence that the core was not being adequately supplied with cooling water—would have been obvious. Case, however, in a 1:40 p.m. briefing of Acting Chairman Victor Gilinsky, made no mention of boiling in the reactor core or the possibility that the core was uncovered. His upbeat report concluded that at the plant, "They're reaching the point where things will get stable in the primary system." Meanwhile, technical analysts who were with Case at the Incident Response Center began to check the reported temperature and pressure data describing conditions in the TMI-2 reactor. They used the standard "steam tables" and noted that there was an apparent superheated steam condition. Victor Stello, one of the key members of the technical staff at the Incident Response Center, believed the data indicated that the core might be uncovered and badly damaged. (Shortly before reaching this conclusion, Stello had talked over the phone with a congressional aide and had discounted the high radiation readings in the containment building and described the reactor as full of water.) Stello reported his disturbing new findings directly to Acting Chairman Gilinsky later that afternoon. According to the transcript that has been made of this telephone conversation, which took place around 4:45 p.m., Stello told Gilinsky that "the only plausible explanation" of the data from the plant, assuming that the temperature readings were correct, was "superheating." To this report, Gilinsky, a physicist, replied, "Well, let me understand what you're saying. You're saying that, in fact, the core may not be covered." Stello replied, "Right." Other N.R.C. analysts preferred to think that perhaps the temperature readings were incorrect. Their view rather than Stello's prevailed, and the N.R.C. adopted an optimistic official prognosis for the situation at TMI-2.

Harold Denton, who finally went to the Incident Response Center at the end of his workday—"around quitting time, at five or so," he recalls—says, "It was at that time that we still thought we could explain the radiation levels from iodine spiking [a phenomenon that sometimes accompanies normal plant shutdown], cladding perforations [small holes in some of the fuel rods], and so forth." Commissioner John Ahearne, a former Defense Department physicist who had only been appointed to the N.R.C. a few months prior to the accident, went to the Incident Response Center as an observer that morning and stayed until well past midnight. He says that he never got the impression during the day from N.R.C. analysts that there had been serious damage to the reactor core. It was not, in fact, until Friday morning—two days after the accident began—that N.R.C.'s analysts finally recognized that the core was badly damaged.

At 5 p.m. on Wednesday the 28th, Stello's minority view having been rejected, the N.R.C. issued a press release that made no mention of superheated steam and its possible dire consequences. Nor was there any disclosure of the radiation level in the containment building—twenty thousand rads per hour. Instead, the press release presented a reassuring view of the current state of the plant. N.R.C. press officers, in preparing the news release, had consulted with senior agency officials, whose suggestions were focused more on the public-relations impact of the statement than on its technical accuracy. Commissioner Richard Kennedy, for example, was reluctant even to call the event at the plant an "accident" and tried to substitute a euphemism. He was bluntly told by N.R.C.'s public-relations chief Joseph Fouchard that there was no way that the term "accident" could be avoided. Officials tried very hard, nevertheless, to phrase the news release in a way that would arouse as little public concern as possible about the leakage of radioactive materials from the plant. The statement indicated that the low levels of radiation measured in the vicinity of TMI-2—it mentioned a "maximum" reading of three millirems per hour—were attributable to small amounts of radiation "shining" through the walls of the containment building rather than to any release of radioactive materials into the environment. This was a technically incorrect and misleading statement, according to N.R.C.'s own subsequent investigation of the accident. There had, in fact, been leaks during the day, with a dose of seventy millirems per hour reported at the north gate of the plant. Moreover, some of the experienced reporters covering the accident knew that the walls of the containment building were so thick that radiation could penetrate them only if there was an astronomical amount of radioactive material inside the building. The N.R.C.'s attempt to downplay the severity of the accident backfired. Subsequent news stories broadcast by the national television networks and the wire services noted the implied high radiation levels in the containment building, and therefore did not convey the favorable impression of the situation that had been intended.

One of the Incident Response Center's functions is to provide information to federal agencies and departments that might be called upon to assist in the response to nuclear-plant accidents—for example, by helping to organize medical treatment for the victims or housing for evacuees. Accordingly, Bernard Weiss, an N.R.C. official in the response center, was directed by senior N.R.C. officials late on Wednesday afternoon to brief other government agencies on the status of the plant. According to the transcripts of his telephone calls, Weiss conveyed the same optimism as the N.R.C. press release prepared for more public distribution. "It was really never a problem with regard to loss of water and exposure of the core," he told officials of the Department of Health, Education, and Welfare shortly before 5 p.m. In a similar conversation with the White House Situation Room an hour later, he reported that "there was never a problem with regard to keeping the core covered."

The reluctance of senior N.R.C. officials to admit the severity of the accident, both to themselves and to the general public, had very important consequences for public safety. As a result of the sanguine official attitude, no consideration was given to the advisability of a precautionary evacuation of the surrounding population. "I don't think anyone raised with me on that day that conditions were such as to warrant evacuation. I don't think I ever considered it myself," Harold Denton acknowledges. Yet, according to N.R.C.'s subsequent investigation of its handling of the accident, there is little doubt that such an evacuation would have been "prudent," given the precarious conditions at the plant that morning. According to N.R.C. Commissioner Peter Bradford:

> If the seriousness of possible nuclear plant accidents were rated on a scale of one to ten, with five being the threshold for recommending an evacuation, N.R.C. on Wednesday thought the accident was a "two" when in fact it was closer to a "seven." If anybody had told us on Tuesday that the Three Mile Island core was going to be uncovered the next day, I can't imagine that we would have failed to order an evacuation.

By misdiagnosing the accident and underestimating its severity, the N.R.C. allowed hundreds of thousands of people to remain in the surrounding communities while the nearby TMI-2 reactor was perilously out of control.

* * *

Little known to the general public prior to the TMI-2 accident, the Nuclear Regulatory Commission, established by Congress in 1975, is a small, highly specialized bureaucracy squeezed in among the behemoth federal agencies that populate our nation's capital. Its total annual budget, minuscule by our central government's standards, would barely support the Department of Defense for a day, and most of the employees on the agency's tiny technical staff could be housed in a few back rooms at the Department of Health and Human Services. Unlike the large government agencies, the N.R.C. has no monumental stone-and-glass structure as its Washington, D.C., headquarters; it doesn't even have its own building. The five N.R.C. commissioners and some of the agency's senior staff are housed in a few dozen rooms in a drab office building in downtown Washington, which N.R.C. shares with several other government agencies. Its technical staff is relegated to widely separated office buildings in nearby Bethesda, Silver Spring, and Rockville, Maryland. Moving the agency's staff under one roof has been discussed for years, but no concrete proposal has been adopted to do so.

The N.R.C. has been unable to gain in status and prestige what it lacks in size. Appointments to the five-member commission are not the most sought-after posts in Washington, partly because the agency's work is considered difficult and technical, and partly because of the many unpleasant controversies in which the N.R.C. has become involved. The N.R.C.'s work is seldom celebrated on the front pages of the newspapers or on the television evening news programs. Its successes are easily overlooked, and what little time it spends in the limelight is usually devoted to answering questions about why some unexpected safety problem has arisen in a nuclear power plant operating under its nominal supervision. The growing controversy over nuclear safety has subjected the N.R.C. to attacks on all fronts, but the most trying problems for the agency have been created by widely publicized charges from some of its own staff members that it has covered up or otherwise mishandled a variety of important reactor-safety issues. For the scientists and engineers on the N.R.C. staff, the constant outside criticism and acrimonious internal controversies have proved quite depressing. One senior N.R.C. official explains:

> Morale is lousy because most of the guys, you see, got into the nuclear program when it was in its heyday. They were the respected and admired "men in the white coats" who would use their slide rules and skills to bring in the atomic-powered age. They were heroes. Nowadays, with all the criticisms and slanders aimed at nuclear power, the social status of "somebody in the nuclear power program" is about on the same level as "murderer" or "rapist."

The chairman of the five-member N.R.C. assumes general administrative duties but has only one vote, like each of the other four commissioners, on policy matters. The commission has been in turmoil practically since the day it opened for business on January 19, 1975. It has had a high attrition rate, and has already seen four chairmen in its brief history. Its first chairman, ex-astronaut and former A.E.C. Commissioner William Anders, left in 1976, a few months after Robert Pollard, one of the N.R.C. staff safety experts, resigned in protest and brought about a major controversy over the agency's safety policies. Anders subsequently went to work for General Electric, one of the principal U.S. nuclear-equipment manufacturers. Another member of the initial group of commissioners, Edward Mason, who came to the agency from the Nuclear Engineering Department at the Massachusetts Institute of Technology, resigned to take a job with Exxon. The second chairman, Marcus Rowden, an attorney, was not reapponted by President Carter when his term expired in June 1977; he now works for a Washington law firm that represents the nuclear industry.

The chairman of N.R.C. at the time of the TMI-2 accident was Dr. Joseph M. Hendrie, a former A.E.C. regulatory staff official and a former chairman of the A.E.C.'s Advisory Committee on Reactor Safeguards. A physicist by training, Hendrie was the only member of the commission with a heavy technical background in nuclear-reactor technology. His nomination as N.R.C. chairman by President Carter was strenuously opposed by some in Congress, such as Representative Morris Udall, who felt that Hendrie's long association with the A.E.C. went counter to the congressional desire to have the N.R.C. take a fresh look at the questions of nuclear safety. Congressman Udall, who headed one of the key congressional committees that oversees the N.R.C., publicly criticized Hendrie as unfit for the post because he was a member of the "old A.E.C. clique." The tall, thin, almost gaunt N.R.C. chairman, who was confirmed by the Senate despite such protests, had to become accustomed to sitting at witness tables in various congressional hearing rooms peering over his half-glasses at congressmen who were often very hostile.

The other four members of the commission when the accident occurred were: Victor Gilinsky, a former RAND Corporation physicist who had also served on the A.E.C. staff; Richard T. Kennedy, a Harvard Business School graduate who came to the N.R.C. from Henry Kissinger's National Security Council staff; Peter A. Bradford, a lawyer whose previous position was as chairman of the Public Utilities Commission in Maine; and John F. Ahearne, a former Defense Department physicist who joined the commission only a few months prior to the accident. Commissioners Bradford and Kennedy have no technical training, and Commissioners Gilinsky and Ahearne, who have the advantage of professional training in physics, have little specific knowledge of the highly specialized field of nuclear-reactor technology. Technical analysis is routinely delegated by the commission to the agency's technical staff; the commissioners themselves seldom exercise independent technical judgment and are generally unable even to understand the staff technical analysis that underlies commission decisions. The five commissioners, in the description of one of them, are the "queen bees" attended by swarming squadrons of technical aides who are given a large measure of discretion in formulating agency policy.

The commission has proved distinctly unable to form a collective judgment on what nuclear-plant safety policies and regulations the federal government ought to adopt. It has sharply divided on just about every major substantive matter that passed before it and, for lack of three votes in favor of any particular policy, has remained largely paralyzed. Chairman Hendrie and Commissioner Kennedy were generally identified as champions of the nuclear industry, and they strongly supported the basic "regulatory framework," emphasizing "industry self-regulation," that Hendrie himself, as a senior A.E.C. official, helped to develop. They maintained that the N.R.C.'s regulations were adequate to protect the health and safety of the public and that, taken together, the regulations provided "defense in depth" against serious reactor accidents. Accordingly, they strongly resisted broad new proposals to tighten the regulations or to curtail the operation of any of the existing plants. Commissioners Bradford and Gilinsky, in contrast, while not "anti" nuclear power, have been severely critical of important agency policies and see the need for a major overhaul of the N.R.C.'s basic safety and licensing regulations. Bradford, with Gilinsky's assistance, has put together what he calls a "list of nuclear self-delusions": his own private catalogue of the mistaken premises that have shaped the development of A.E.C. and N.R.C. policies on nuclear power. According to one item on this puckish commissioner's list, the N.R.C. is quick to dismiss new technical evidence of reactor-safety flaws because its long-standing rationalization has been that "no chain is weaker than its strongest link."

Relations between Bradford and Gilinsky on one "side,"

and Hendrie and Kennedy on the other, were not always cordial, and divergent outlooks have turned some of the commission's nominal decision-making meetings into endless sparring sessions. Caught somewhere in the middle was the fifth and most recently appointed commissioner, John Ahearne, who often cast the "swing vote" on the commission. He has generally swung toward positions urged by the N.R.C. staff, which did not lead necessarily to any coherent policy, since the staff itself has yet to develop concrete policies on many issues and is awaiting guidance from the commission. "What is the goddamn position of this agency on anything?" one senior member of the N.R.C. staff asked in an interview a few months after the accident. "What are our marching orders from the commission? What regulations are we supposed to enforce? What level of safety are we supposed to be promoting? How hard are we supposed to push the industry? Or do they just want us to throw in the towel?"

Paralyzed by controversy and uncertainty, the commissioners, according to one industry newsletter, have pondered the agency's business "like five Hamlets." In the meantime, unresolved questions about their safety notwithstanding, some six dozen licensed nuclear-power plants remain in operation.

* * *

Having officially discounted the risks of serious reactor accidents, the N.R.C. had never attached a high priority to the development of contingency plans for handling such an event as the TMI-2 crisis. The N.R.C. staff was not adequately trained for this purpose. No special communications links were established so that the N.R.C. would quickly get the technical information needed to understand conditions at an affected plant. No emergency decision-making procedures had been thought through and established. The N.R.C. Incident Response Center was a token emergency headquarters that had been set up largely as a result of the 1975 accident at Browns Ferry Nuclear Power Station near Decatur, Alabama. Plans for an N.R.C. emergency center were set forth in an internal report dated July 23, 1976, prepared by Brian Grimes of the N.R.C. staff. The memo outlined the personnel, communications systems, technical information, computer assistance, and other procedural and logistical requirements for an effective program to assure that the N.R.C. competently monitored, advised, and, if need be, directed nuclear-plant operators during serious reactor accidents. The N.R.C. adopted the recommendations in his report "in principle," Grimes said in an interview, but never implemented them in any detail.

Thus, the Incident Response Center, as it was finally set up, was an austere command post that consisted of a few rooms equipped only with desks, telephones, a few blackboards and maps, and folders for each operating plant containing the facility's emergency telephone numbers and a description of its emergency plans and procedures. The center was not provided, as Grimes had recommended, with detailed, "precollected" technical data about each plant, information that N.R.C. personnel might need to consult during an emergency, nor did the center even contain a general library of basic technical reference works. There were copies in the Incident Response Center of the Final Safety Analysis reports for "some" of the operating nuclear plants, Grimes stated. These are the official documents, submitted to the N.R.C. by the owners of each nuclear plant, that provide technical details about the plant's reactor and safety-systems designs. Upon arriving at the Incident Response Center during the accident, N.R.C. officials, according to Grimes, discovered that the Final Safety Analysis Report for TMI-2 was missing.

The conventional telephones in the response center were not part of any special communications network, although one of the emphatic recommendations of the Grimes memorandum was for "a system of communications which does not rely on the telephone network." An N.R.C. study group reviewing the Browns Ferry accident had also specifically called attention to the need to get better communications arrangements, and the N.R.C. subsequently hired a consulting firm, the MITRE Corporation, to review and assess the kinds of communications links that might be used. The report from the MITRE Corporation was completed in November 1977, but none of the special communications systems it discussed had been put in place by March 28, 1979. The N.R.C.'s response center, as a result, had no hotlines or other forms of dedicated connections to assure a reliable flow of information between the response center and the TMI-2 control room.

One of the recommendations in the Grimes report was for a "hookup" between a computer in the N.R.C. response center and the computers in plant control rooms so that the N.R.C., aided by modern data-transmission technology, might have quick access to all vital data about what was happening at the plant. Data from the plant could then be displayed on an instrument panel at the Incident Response Center, thereby allowing N.R.C. officials to monitor the same basic data as the reactor operators. This recommendation was also ignored and no computers or any other type of sophisticated equipment was provided to transmit information from nuclear plants to the N.R.C. Even the simple recommendation that N.R.C. response-center personnel have formal "checklists" available—to make sure they systematically monitored a troubled reactor's vital signs and asked all the right questions—was not heeded. Lacking a basic checklist, Grimes noted, N.R.C. officials forgot to ask some important questions during the TMI-2 accident.

Telephone communications between the N.R.C. Incident Response Center and the TMI-2 control room proved extremely difficult during the accident. Part of the time the N.R.C. headquarters could only communicate with the plant indirectly: officials in Bethesda would call the agency's regional office near Philadelphia, and officials there, on another line, would call the TMI-2 control room. At times N.R.C. headquarters had to communicate through the TMI-1 control room, which had another line to the TMI-2 control room. There was a period when all telephone communications between N.R.C. and the plant failed completely. At other times, when the N.R.C. was depending on the Unit 1 control room to relay information, communications between Unit 1 and Unit 2 failed, and the N.R.C. had to wait while couriers were sent from one building to another, a task that became increasingly difficult during the day when airborne radioactive materials required the messengers to don special protective clothing, face masks, and respirators (which resembled

SCUBA tanks). For much of the day, the TMI-2 control room personnel themselves had to wear face masks and could only talk to one another by shouting or, at their own personal risk, by removing the face masks, as they frequently did when using the telephone. The company officials to whom the N.R.C. talked, moreover, were relatively junior employees who were not involved in the decisions that were being made and who were often not even briefed on plant conditions by their superiors, those senior officials being too busy to talk with anybody.

As problematic as communications were between the Incident Response Center and the plant, this difficulty ultimately proved to be only a subsidiary contributor to the N.R.C.'s initial misdiagnosis of the problem inside the TMI-2 reactor. Despite the awkward and erratic telephone links, N.R.C. officials say they were able to get most of the basic information from the plant that they requested. According to Victor Stello:

> We had a line directly when I got there with a guy in the control room getting whatever information we were looking for. . . . Whatever information I wanted, he would go get it and give you whatever else that you didn't ask for that was developing up there. . . . I didn't perceive any lack of getting the technical information we were trying to get in trying to decide what was going on in the core. The difficulty, of course, is being able to digest, understand, evaluate all of what I'd heard.

Central to the Incident Response Center's difficulties in diagnosing the accident was the N.R.C.'s lack of high-level technical competence. No one on the N.R.C. staff was familiar enough with TMI-2 or other Babcock & Wilcox reactors to give specific guidance or directives to the operators of the stricken plant. Such practical competence among its staff has not been an N.R.C. priority. Nor is advancement within the agency closely linked with technical, as opposed to bureaucratic, skills. The relatively small group of N.R.C. senior officials, some of whom were members of the A.E.C. regulatory staff in the early 1960s, are career bureaucrats far more familiar with the protocols of their closed agency world than with the subtle technical issues affecting nuclear-reactor safety. Harold Denton, for example, is a cheerful, friendly, almost courtly electrical engineer who rose through the ranks in his career with the A.E.C. and N.R.C. He said in a recent interview that at the time of the TMI-2 accident, he had "no more than a passing acquaintance" with the specific design features of Babcock & Wilcox nuclear plants. Prior to the Three Mile Island accident, Denton explained, he had been to other operating Babcock & Wilcox plants (or at least he thought he had), and he remembered that he "had gotten the normal Cook's tour" of Babcock & Wilcox facilities at the company's headquarters in Lynchburg, Virginia. Of course, in making his assessment of conditions at TMI-2 during the accident, Denton could call on other agency officials with more specific knowledge. Yet according to one of Denton's senior aides, "There is a tremendous element of the 'buddy system' in this organization, and the 'old cast of characters' rather than those with specific knowledge of Babcock & Wilcox reactors were the ones who called the shots during the accident."

In addition to the disadvantage of limited technical competence, N.R.C. analysts were also hindered in their diagnosis of conditions at TMI-2 by the agency's rigid presuppositions about the nature of serious reactor accidents. The N.R.C carries out its official safety analyses under a set of rules that deem certain types of accidents "credible" and others "incredible." Limiting nuclear-plant-safety requirements to protection against a relatively narrow set of "credible" accidents that are to be controlled by select "safety" equipment was done on an ad hoc basis without any systematic probability analysis as justification. Indeed, in 1975, when the N.R.C. finally completed an elaborate accident-probability study begun by the A.E.C. in 1972, one of the principal findings was that certain allegedly "low-probability" accidents, ruled "incredible" when plants were given their official licensing reviews, in fact had higher probabilities of producing meltdown accidents than the accidents deemed "credible" by the N.R.C. N.R.C. safety and licensing reviews, in other words, according to the agency's *Reactor Safety Study,* were ignoring what might be the dominant sources of nuclear-power-plant accident risks—such as small loss-of-coolant accidents (a problem that can be created by a relief valve that jams open). Dr. Stephen Hanauer noted this general deficiency in the agency's safety analyses in a 1975 internal memo sent directly to the N.R.C. commissioners entitled "Important Technical Reactor Safety Issues Facing the Commission Now or in the Near Future." Dr. Hanauer wrote:

> The study has pointed out a disparity between a) our present . . . safety approach in which all potential accidents are either put into the design basis for complete mitigation or remain outside the design basis and have no safeguards compared to b) the more realistic viewpoint of a spectrum of accidents each with probability and consequences of its own. Serious consideration should be given to modifying the present all-or-nothing approach in the light of reality.

Dr. Hanauer's advice was not followed, and the orthodox safety-analysis philosophy of the A.E.C., aimed at minimizing nuclear-plant-construction costs, continued in use by the N.R.C. This "decision"—which was made by default, Hanauer's memorandum simply having been ignored—had profound repercussions on the design of plants like Three Mile Island, which were, of course, unprotected against multiple failures and all other types of allegedly "incredible" accidents. The decision also meant that N.R.C. safety analysts would be forced to think only in terms of a well-ordered, antiseptic, ideal universe in which only certain carefully designated things are allowed to go wrong. Moreover, according to the official way of thinking, for each "accident," there was always appropriate "safety" equipment available, which was assumed to perform satisfactorily. Like the reactor operators, trained on control-room simulators in which everything more or less works according to plan, N.R.C. officials were ill-prepared for the messy problem they faced on March 28, 1979.

N.R.C. officials, according to Harold Denton, always had the thought "in the back of our minds" that some unlucky plant on some unlucky day might have a serious accident. The accident they envisioned was a brusque, dramatic, easily recognizable event, something as easily comprehended as a plane crash or a bridge collapse. As they conceived of the event, Denton says, it would be a "maximum credible accident," the most severe type of Class Eight accident on the N.R.C.'s highly selective list of hypothetical accidents, and it would be over and done with *before* the N.R.C. could get involved. Denton noted:

> Any incident response up to this time had been on the presumption that accidents would happen all in a hurry and our role was sort of monitoring, staying on top of what was going on, and reporting back to civil defense authorities and so forth, but not one to be an active participant, so to speak. My own perception of what the Incident Center did in the way it was set up, was to have information flow in, but not direct or control. We would hear what was happening. I think the thought was that accidents like a large [loss-of-coolant accident] happen all so fast. If the emergency core-cooling system would all work, the whole thing would be over within an hour; it either worked or it didn't work. That's my kind of perception. What we would do would be to call and coordinate federal authorities, and so forth. I think, at least, I didn't anticipate getting involved in controlling . . . day-to-day changes that were being made at the plant.

Commissioner Ahearne, as an observer at the Incident Response Center, found that:

> People there believed they were in a monitoring role and they had confidence that procedures in place [at TMI-2] were adequate for coping with the accident. N.R.C.'s relationship is such that the licensees have the responsibilities during accidents. N.R.C.'s role is to follow what is happening. If it looks like the licensee is going to take some action that is wrong, then N.R.C. would step in. I was puzzled why N.R.C. didn't take a more direct role. But I guess it is not N.R.C.'s style of operation to run the show.

* * *

There was no question on that Wednesday morning in the N.R.C. Incident Response Center, or in Gary Miller's mind, who was running the show at Three Mile Island. "It was clear that it was my decision," says Metropolitan Edison Company's station manager Miller, who was in charge of operations and maintenance for both units of Three Mile Island Nuclear Power Station. Miller, after various telephone conversations with plant officials, arrived at TMI-2 just after 7 a.m. A "site emergency" had already been declared, and the TMI-2 control room, filled at times with as many as sixty people, was a noisy and confused command center. "I'm not trying to be funny about it," operator Craig Faust has recalled, "but it was a little tense in there." Miller formally took charge, announcing that he was the emergency director. He appears to have been calm and, at least initially, well organized and decisive, his first major act being the declaration at 7:24 a.m. of the country's first nuclear-power-plant General Emergency. This was the beginning of what Miller later described as his "ten hours of hell."

He received a quick briefing from plant personnel and, not wanting to lose any time in executing prearranged plant emergency procedures, he cleared the control room of unnecessary personnel. He recalls:

> We run emergency drills every year, and I run those drills. When I got to the control room, it was 7:15 in the morning, the radiation indicators were escalating—that's the best way to put it—and they were on everywhere. I just took one guy at a time, and I said, "You're in charge of that" . . . I took one guy and I put him in charge of reading the emergency plan. I took one other guy and put him in charge of making all the calls—another guy in charge of the technical support.

Miller's highest priority was organizing teams to monitor offsite radiation levels to determine whether radioactive materials in any dangerous amounts might be escaping from the plant. "The biggest thing is getting teams out there with radiation monitors," he asserted.

Deploying teams to measure radiation levels in the vicinity of the plant ought to have been a straightforward task. Radiation-monitoring equipment is commercially available, and the plant had standard written procedures governing the use of these devices in an emergency. One of the pertinent procedures contained the plant's "Emergency Readiness Check List," which indicated that four "Radiation Emergency Kits" were supposed to be available at the plant, each consisting of radiation-monitoring equipment, containers for air and water samples, and other useful items. Only three of the four TMI-2 Radiation Emergency Kits were stored in their proper location (the fourth was in a supervisor's office because some of its equipment was inoperable; the equipment had been out of service for weeks before the accident but had not been repaired or replaced). When, prior to leaving the site, plant employees made a brief check of the equipment in the other three kits, they found more radiation detectors that were not functional. Once the radiation teams were in the field, further problems with the equipment developed. The principal device required for the initial radiation surveys was an instrument known as a PIC-6A. TMI-2 had a nominal supply of fourteen of these devices, of which only four were available in working condition at the time of the accident.

Defective equipment was not the only problem. The employees assigned to the radiation-monitoring teams had not been adequately trained for their assignment. Some of them tried to use the monitoring devices without taking off the cover that shielded the radiation-sensitive part of the instrument, and most of them failed to record the readings properly so they could be used to assess potential radiation doses to the population. Some of the technicians couldn't find the vehicles they were supposed to use to transport themselves and their equipment to their assigned locations. Teams were also delayed in reaching nearby communities on the west bank of the Susquehanna River because of the lack of bridges. Goldsboro, for example, is only about a mile and a half away from the plant, but by car that morning it was more than a twenty-mile drive. The first team to arrive in Goldsboro, moreover, discovered that their emergency kit contained inoperable radiation-monitoring equipment. All of these problems would have been obviated if permanent radiation-monitoring equipment had been installed in the communities around the plant, equipment that would automatically report to the control room the radiation levels in key locations.

The off-site radiation levels reported by the survey teams

around 8 a.m. indicated dose rates of only a few millirems per hour. (The standard unit for describing radiation dosage is the rem, and doses are generally measured in terms of the derivative unit, the millirem, which is one one-thousandth of a rem.) Plant officials considered the dose rates to be quite low. They were, in fact, well above normal "background" radiation levels—the amount of radiation we all receive from cosmic rays and the trace quantities of radioactive materials normally present in rocks, soil, water, and building materials, and in our own bodies—but still well below the levels that most experts would associate with serious health effects, presuming, of course, that exposure is brief. Miller concluded that while the radiation releases from the plant had to be stopped, there was no necessity for an immediate evacuation of the neighboring population. Miller's own ten-year-old daughter, he noted in a later discussion of his decision, was at home at this time, a few miles away from the plant.

* * *

Organizing the parts of the plant-emergency procedures relating to notification of outside agencies and off-site radiation monitoring, Miller has acknowledged, was a straightforward, preplanned, almost mechanical task that called for little exercise of judgment. By far the more challenging problem he and other senior plant officials faced that morning was the development of a plan to stabilize the reactor. Miller suspected that there had been some damage to the reactor fuel for the simple reason that the fission products, the radioactive debris created by the nuclear chain reactions, are normally contained inside the fuel rods, and some form of damage had to have occurred for the radioactive materials to get out of the fuel rods and into the containment chamber. He initially believed that "crud burst" or minor fuel-rod cracks might account for the radiation alarms. In addition to these possibilities, Miller has explained, he thought that some of the high radiation levels and high temperature readings on plant instruments might be the result of malfunctioning instruments. It's "not unusual to see instrument error," he has said, an observation that can be supported by any casual survey of nuclear-power-plant instrument performance. Therefore, he tended to attribute the extremely high readings from the radiation monitor in the upper dome of the containment chamber to a defect in this instrument. He believed that *some* radiation was present, a fact that was consistent with other radiation-level indicators in other parts of the plant, but he refused to believe what he thought were impossibly high readings from this device.

In fact, there was a serious defect in the arrangement that was used to monitor the radiation level in the dome of the containment building, and it accounts for much of Miller's confusion, as well as for the N.R.C.'s failure to believe the urgent warning signs provided by this instrument. The device in question, which is installed under the roof of the building housing the reactor, consists of a radiation monitor that is surrounded by two inches of lead shielding. As a result of the shielding, only a small fraction of the radiation in the building penetrates the lead and is actually detected by the monitoring instrument. (Since the device was installed to measure radiation levels during accidents, the designers apparently didn't want it to be picking up the small amounts of radiation that might be present in the containment building during normal operation.) Plant officials believed at the time that the lead shielding provided a one-hundredfold "attenuation factor"—in other words, that the actual radiation levels in the building were one hundred times higher than reported by the instrument. Thus, when the device read "200 rems"—as it did that morning—they interpreted this "raw data" in the conventional manner and inferred that a radiation level of twenty thousand rems was being indicated by the instrument, which is what they reported to the N.R.C. They knew, however, that such a radiation level was ridiculously high, far higher than would ever occur under the worst conceivable accident, and so concluded that the device was simply malfunctioning. What they did not know either in the control room or at the N.R.C. was that the lead shielding had not been installed perfectly in one solid piece, so holes allowed the monitoring device to pick up more radiation than expected. The standard one-hundredfold "attenuation factor" was incorrect, and some still-not-precisely-determined lesser factor ought to have been used in interpreting the readings from the monitor. Once again, plant-instrumentation deficiencies were "keeping secrets" from the personnel responsible for crisis management.

Another major instrument problem that Miller confronted involved the temperature-measuring devices known as thermocouples that were installed in the reactor, just above the core, to monitor the temperatures in the reactor during normal operation. Readings from the thermocouples had to be obtained from the computer in the control room, which only gave temperature readings up to seven hundred degrees; when the temperature got any higher, the computer just printed out a string of question marks. Miller asked Ivan Porter, TMI-2's instrumentation-control engineer, whether there was another way, besides the computer, to find out what temperatures the thermocouples were really measuring. Porter told him that it might be possible to find the wires leading from the thermocouples in the reactor back to the computer and to attach a measuring device right on the wires to pick up the actual temperature of the reactor, the way one might "tap" a telephone line and pick up the conversation transmitted. Porter put together a team of instrument technicians who looked through plant-instrument drawings and located the correct wires leading back to the thermocouples in a room directly below the control room. The technicians took a series of readings that shocked them, they have said, because they indicated that the core was uncovered and overheating badly. Two of the readings they reported to Porter indicated temperatures of approximately twenty-three hundred degrees, spectacularly higher than the normal six hundred degrees. Some of the readings, however, were very low—only about two hundred degrees. Porter did not think the low readings were believable, since there was other evidence of temperatures above seven hundred degrees in the pipes coming out of the reactor, and he similarly dismissed the higher temperature readings as unreliable. The instrument technicians, however, have stated that to overcome Porter's doubts, they reconfirmed the higher temperature readings in his presence using another instrument. Still, Porter would not accept the evidence. "I believe Ivan didn't want to believe what was taking place," Roy Yeager, one of the instrument technicians explained, adding:

> I don't know whether it was an attitude of "Hey, your instruments are wrong—you guys don't know what the heck you're doing"—or what-not. I think the general consensus [at TMI-2] throughout the whole first day was, number one, nobody really knew what was actually happening; number two, some that had an inkling of what was happening didn't really want to believe what was going on.

When Porter reported the readings to Gary Miller in the control room that Wednesday morning, he told Miller that he did not have confidence in their accuracy. Miller himself has said that he also did not believe the high temperature readings. He saw no consistency in the data—one would expect the temperature readings, since they came from closely spaced measuring devices, to show some general agreement—and he distrusted the instruments themselves, since they were not standard equipment specifically "qualified" and tested to establish their reliability under accident conditions.

After Porter left them to report the data to Miller the instrument technicians continued to take additional readings and obtained similar results. At about this time, the technicians themselves were evacuated from Unit 2, along with other "nonessential personnel," because of the high radiation readings in the plant. Both Porter and Miller said that they did not see the additional data, nor did Miller—or anyone else at TMI-2—inform the N.R.C. that day about any of these special temperature readings. The N.R.C. did not, in fact, learn about these data until weeks after the accident.

By midmorning, having done all the simpler things called for in plant-emergency procedures—informing outside authorities, dispatching radiation-monitoring teams, and the like—Miller still did not know what was happening or just what he was supposed to do about it. Untrained in the handling of such a complex emergency, he "never imagined," he has said, that the core was uncovered and that ferocious chemical reactions had already destroyed much of it. The plant engineering staff was by then available and at his disposal, and he had organized his chief lieutenants into a "senior think tank" that met several times an hour to review plant conditions. They, too, were stymied by the same instrument deficiencies and lack of technical skills that had frustrated the operators hours before, and they were unable to diagnose plant conditions or develop any remedial strategy. "I was grasping at straws trying to assess what was happening," senior plant engineer George Kunder recalled. Everyone recognized that there was a serious problem, Miller has explained, but no one had the solution. Hours passed, and the TMI-2 reactor, for want of adequate cooling, remained near the brink of a core meltdown.

* * *

Miller and other members of his staff were not left entirely on their own at the plant in the effort to bring the TMI-2 reactor back under control. From the late morning on, the N.R.C. had a direct presence in the TMI-2 control room and in other locations around the plant. At 8:40 a.m., the agency's regional office near Philadelphia, which is about eighty-five miles away, had dispatched a station wagon full of N.R.C. officials, who arrived at the plant shortly after ten o'clock. The team consisted of two N.R.C. inspectors and three health physicists (radiation-monitoring specialists). An hour later, two more N.R.C. health physicists arrived on the site. The N.R.C. team did not greatly augment the level of technical expertise represented by the Metropolitan Edison staff. Only one of the seven N.R.C. officials had substantial familiarity with TMI-2, and none of them was competent to operate the reactor. The N.R.C. team was unable to give specific emergency instructions and had no explicit instructions from N.R.C. management, except for the general assignment to "monitor" plant conditions. Nor was it even clear who was in charge of the N.R.C. team. Inspector James Higgins, one of the officials sent to the plant, has said that he didn't know who his boss was. Uninstructed and leaderless, the N.R.C. team was also very poorly equipped, since it had not brought with it adequate radiation-monitoring equipment or the protective clothing, such as face masks and breathing apparatus, that was needed because of the high radiation levels in the plant. (At the plant, they were running out of required breathing apparatus.) The team had equipment for taking air samples to check for the presence of radioactive materials in the vicinity of the plant, for example, but they didn't have batteries to operate it and couldn't use it in the field: it had to be plugged into an alternating-current outlet. The N.R.C. regional office did have a mobile laboratory—a special van—that was equipped with radiation-measuring equipment, but at the time of the accident, it was in use at the Millstone Nuclear Power Plant near New London, Connecticut. The van was driven to TMI-2, but it did not reach the plant until around 7:30 p.m. When N.R.C. officials checked the van after its arrival, they found that it had one flat tire and another tire that was badly worn, and that it was unfit for making field surveys of radiation levels. The N.R.C. special review group that investigated the TMI-2 accident incorporated in its final report a reprint of a cartoon showing a carload of N.R.C. inspectors arriving at the gate of a nuclear plant. The cartoon, by Bill Schorr, depicts the inspectors as bug-eyed Keystone Kops riding in an old jalopy, with one of them studying a manual entitled *Your Friend the Atom.*

When the N.R.C. team arrived on the island, they reported to the TMI-1 control room, which had been turned into the headquarters from which the monitoring of off-site radiation levels was being directed. There the inspectors received a general briefing on plant conditions from Metropolitan Edison officials. N.R.C. Inspector James Higgins, a thirty-one-year-old Annapolis graduate and former nuclear-submarine-reactor operator, then made his way to join Gary Miller in the TMI-2 control room, accompanied by Donald Neely, an N.R.C. health physicist, after both had donned face masks, respirators, and special protective clothing. "Around that time, Unit 2 room was being evacuated of unnecessary personnel because of the high airborne radioactivity levels," Higgins said, remembering also "a little quickening of the pulse" as he made his way through the accident-damaged plant. There were twenty or thirty people in the TMI-2 control room when he arrived at about 11 a.m., but he felt that the number of people was not hindering emergency operations. "As a matter of fact, not from a numbers standpoint, but from a standpoint of information that was gotten, enough people were not there or probably more people—correction—the right people were not there with the proper knowledge to evaluate what was going on to determine what should have been done," he said. Higgins found Gary Miller and his senior engineers "unsure how to proceed in this situation." So, it turns out, were Higgins and all the other N.R.C. personnel sent to the site.

Higgins, as an N.R.C. inspector, normally performs routine audits of plant records to determine compliance with N.R.C. regulations. N.R.C. field inspectors, although acquainted with the general principles and some of the narrow details of plant operation, do not carry out sophisticated reactor-safety analyses, nor are they trained to serve as skilled emergency managers who can be dispatched to handle nuclear-plant crises. According to Higgins, inspectors like himself were not thoroughly knowledgeable in individual plant systems, instrument locations and arrangements, plant procedures, and the "very nitty-gritty details" familiar to the plant's operators. "Actually," he candidly told accident investigators, "throughout the first day I was in a learning process myself to update myself as to exactly what equipment they had [at TMI-2]." Higgins also explained, "I would not want to put myself in the situation to be put in the control room and taking control of the situation. I don't feel I'd be capable of doing that."

Higgins's formal role was to help with the N.R.C.'s "monitoring" of plant conditions by reporting data back to the Incident Response Center and by transmitting N.R.C.'s instructions and advice, if any, back to the plant. He and his fellow inspectors, according to N.R.C.'s later analysis of its response to the accident, were "errand boys" whose principal function that day was "answering a host of unrelated questions from a host of different people at Region I or headquarters." Communications from the TMI-2 control room back to N.R.C. headquarters were difficult, of course, owing to the fragile and indirect telephone links set up during the day; the frequent necessity to don face masks in the control room, which made it difficult to talk on the telephone; and the lack of any organized N.R.C. plan, or a standard checklist, for monitoring nuclear-plant conditions during accidents. Higgins was on his own in gathering information in the control room, and the questions with which N.R.C. headquarters besieged him, he said, "would have taken ten people to answer." (His fellow inspectors were carrying out other assignments, such as checking radiation levels around the plant.) N.R.C. officials in the Incident Response Center, he found, were asking many peripheral and irrelevant questions, had great difficulty understanding the information he was passing along to them because of their unfamiliarity with the design details of the plant, and kept losing track of information he had already transmitted. His own unfamiliarity with the sprawling TMI-2 control room contributed to his difficulties in getting the information requested by N.R.C. headquarters. Not knowing where to look for the data on the complicated plant-instrument panels, he had to wait for assistance from the control-room personnel, who had many other pressing responsibilities to attend to. As he went through his own "learning process," he gradually improved his data-gathering abilities, and has explained that "by the end of the day, I was able to answer a lot of questions over the phones by going out and getting the information myself directly from the [instrument] panels."

Higgins disregarded what he thought were some of the more pointless questions from the Incident Response Center and tried to use some of his time to make his own review of the main problems facing the plant, he explained in a recent interview. He found, however, that so many unexpected problems and incidental difficulties kept arising at the plant that it was difficult for him, as it was for Miller and his crew, to do any real engineering analysis. Higgins was not kept very busy, however, by his nominal role as middleman transmitting advice and guidance to the TMI-2 staff from the Incident Response Center. "Quite frankly," he said, "Washington was not much help that day." Even when Incident Response Center officials had determined that superheated-steam conditions might have been present, and that the core was possibly uncovered, they issued no directive to the plant staff to rectify the problem.

One of the few exceptions was a single conversation that Victor Stello had around four o'clock with Gregory Hitz, a shift supervisor in the TMI-1 control room. Stello, as he has explained it, "impulsively" grabbed the telephone to relay his concern about the superheated-steam conditions. He said to Hitz, who had been talking with response center personnel on another matter, "Let me bounce a question off you." Stello continued, according to the transcripts of this conversation, "If you really have a five hundred fifty degrees on that hot leg"—one of the pipes coming out of the reactor—"it's clear that you're getting some superheat. If you're getting superheat, there's a chance the core could be uncovered.... Have you thought about what problem you've got, if indeed you've got five hundred fifty degrees on that hot leg ...?" Hitz explained why plant officials thought the core was covered but said that he would pass on Stello's concern to the TMI-2 control room. Stello then directed Norman Moseley in the Incident Response Center to ask Inspector Higgins for an explanation of the high temperature readings that Stello believed indicated superheated steam. Higgins later said that he had no recollection of this phone call and did not learn of Stello's concerns that day, but the response center's tapes show that this call did take place. (Since the center was asking for an "explanation" of the data, one of many such calls that Higgins fielded that day, rather than conveying an official "concern" or directive, the slight impression the call made on Higgins is perhaps understandable.) Higgins, according to the transcript of the conversation, repeated the same arguments that Hitz had just given Stello for believing that the core was covered. Moseley, with Stello standing next to him and listening to the conversation, told Higgins that his reply was "not convincing."

Stello took no further steps, however, after this unsatisfactory conversation with the N.R.C.'s "errand boy" in the TMI-2 control room, to contact the appropriate people at TMI-2 to convey his concerns. "It didn't occur to me to talk to anyone in the management of the plant beyond the individual I talked to," Stello said later, referring to his conversation with Gregory Hitz. He did, however, talk with Commissioner Gilinsky shortly after Moseley's conversation with Higgins. Stello told Gilinsky that he couldn't understand what was being done at the plant to keep the reactor cooled. Gilinsky asked whether the plant staff was technically competent; Stello said that he didn't know. Gilinsky asked, "Who is in charge?"; Stello replied, "Don't know." "Well," Gilinsky asked, "is there anything we ought to do about that, beyond having talked with this guy [Hitz]?" Stello explained that he didn't have enough information to do more than raise questions for plant officials to consider. "We've got to be careful that, you know, they don't start asking us what to do," Gilinsky noted, and Stello quickly agreed. "No," Stello said, "they're in charge, and we can only offer something that we thought of, but they are absolutely in charge. There can't be any question about that. And we don't want any confusion in anybody's mind, especially in their mind. I mean they are on top of the situation and we're at the other end of the telephone."

* * *

The TMI-2 staff was ultimately able to prevent a meltdown of the TMI-2 reactor, though they accomplished this largely through inadvertence. At 6:18 a.m., although the operators did not know the relief valve had been stuck open, the block valve was closed and the leak terminated. This temporarily headed off an incipient meltdown that was, at most, an hour away, according to the later engineering analysis. With the block valve closed, the pressure in the reactor increased, and this had the beneficial effect of collapsing some of the steam bubbles in the system so that the water level in the core increased. (The operators should also have turned on the emergency cooling pumps at this point but failed to do so.) Then, around 8 a.m., after initially directing that the emergency cooling system be kept off, Miller decided to countermand this order and have the pumps turned on. The operators did so, and the resulting water supply, while not enough to provide sufficient core cooling, was enough to head off the uncontrolled overheating of the fuel. Once again, this key action by plant officials reflected a large measure of good luck, for in giving his instruction to the operators, Miller appears not to have specified the rate at which the pumps should operate. Shortly after receiving Miller's directive, the operators increased the rate substantially but still left the pumps "throttled" and operating far below their capacity. At various times later in the day, however, they cut back the rate once again. An entry in a control-room operator's log at 1:10 p.m., for example, noted that the emergency cooling pumps were "stopped." There is no exact information available on how much emergency cooling water was actually pumped into the reactor on the first day of the accident, since no recording device was installed in the TMI-2 control room to preserve this important data. Moreover, the operators, early in the accident, had opened a drain line to remove water from the reactor, in the mistaken belief that there was too much water in it, and they continued to drain water from the reactor throughout the day. No one appears to have paid attention to the net amount of water being delivered to the reactor: the difference between what the emergency cooling pumps were adding and what the open drain line was taking away. That the emergency cooling pumps actually managed to supply a large enough net addition of water to keep the core cooled while the TMI-2 staff deliberated further actions during the next several hours appears to have been largely happenstance.

Miller and his "senior think tank," which included N.R.C. Inspector Higgins, met several times an hour to review the status of the plant and to try to formulate a strategy for stabilizing the reactor. The decision making on that day on how to terminate the accident was confused and erratic. Detailed chronologies of operator actions show a pattern of vacillating activity. The same equipment was turned on and off frequently, the field of action was changed suddenly from one set of equipment to another, and the strategies employed during the day varied markedly from one period to the next. The reasoning behind some of the shifting strategies is often far from evident. At one stage, Miller and his staff wanted to keep the reactor at high pressure so they could restart the main reactor coolant pumps (which only work when the pressure is high enough to prevent steam bubbles from forming in the reactor's cooling water). Yet, when they restored the high pressure in the reactor, they did not try to start the main reactor coolant pumps but scrapped their plan and started to lower the pressure in the reactor. Then after several hours with the reactor kept at low pressure, they reverted to high pressure once again. As Miller himself later admitted, the plant was in a difficult situation and "I didn't know how to get out of it." With Higgins's concurrence—and with the knowledge of the Incident Response Center—the TMI-2 staff shifted to yet another cooling strategy. At around 11:30 a.m., they decided to lower the pressure in the reactor so that the shutdown cooling equipment could be used, equipment that worked only when the reactor was at low pressure. To lower the pressure, they decided to open up the block valve in order to drain cooling water out of the reactor through the relief valve. In effect, they were duplicating the event that started the crisis! Normally, of course, the way to lower the pressure in the reactor is to cool it down, the same way one lowers the pressure inside a pressure cooker by pouring on some cold water. Cooling the reactor, however, was precisely what they were unable to do. And so, in desperation, they decided to repeat the initial phase of the accident (caused by a stuck-open relief valve), in the hope that this second time around, there might be a happy ending: once the pressure fell down to about four hundred psi—it was about two thousand psi when they started the maneuver—they hoped that the shutdown-heat-removal system, a normal method for cooling the reactor, could be turned on.

In retrospect, the decision to lower the pressure inside the reactor was a major technical blunder. To begin with, TMI-2 officials overlooked the fact that the temperature inside the reactor was so high, and there was so much steam being generated inside the overheated core, that it would not be possible within any reasonable period of time to lower the pressure sufficiently so that the shutdown cooling system could be used. The operators never succeeded in getting the pressure much lower than about four hundred and fifty psi, yet they persisted for some six hours in the effort to drain the reactor of its cooling water. Second, by draining the cooling water from the reactor, they jeopardized the cooling of the core, which was already badly damaged and deprived of adequate cooling water, subjecting a core that had barely survived an inadvertent loss-of-coolant accident to another, intentional, loss-of-coolant accident.

One of the reasons the TMI-2 officials undertook such an ill-conceived maneuver was that they thought the core was completely covered with cooling water. According to Higgins, Miller repeatedly asked members of his "senior think tank," including Higgins himself, whether any of them felt that the core was uncovered. No one, said Higgins, thought that it was. Moreover, they discussed a kind of "test" they would perform during the depressurization procedure that they thought would provide additional evidence that the core was covered. Part of the emergency cooling system at the plant consisted of emergency cooling tanks that squirt water into the reactor when the pressure in the reactor falls below six hundred psi. The tanks were like giant aerosol cans: partly filled with water, they used nitrogen gas at a pressure of six hundred psi as a propellant. Thus, when the pressure in the reactor fell below that level, the gas pressure was high enough to push emergency cooling water into the reactor. TMI-2 officials decided that they would monitor the water level in these tanks, and if the tanks discharged only a small amount of water into the reactor, they believed this would be a "positive indication," as Higgins described it, that the reactor already had enough water to cover the core.

Actually, the "senior think tank" misunderstood how these emergency cooling tanks worked. The discharge of water from the tanks was not controlled by the water level in the reactor, but by the pressure of the nitrogen gas relative to the pressure in the reactor. Only if the pressure in the reactor fell sharply below six hundred psi would there be a proportionally large discharge of water. (As water is discharged, the nitrogen gas expands and its pressure, the driving force behind the water, is reduced, in accordance with Boyle's law.) The small discharge of water from the tanks was no proof whatsoever that the core was covered. The "test," Inspector Higgins acknowledged in an interview, "was not the best way to go."

Interestingly enough, like some of their other inadvertent actions, the depressurization blunder had its positive aspects. The failure itself was quite fortuitous, because if the operators had actually succeeded in using the shutdown cooling system, there might have been very large leaks of radioactive materials from the plant. The shutdown cooling system works by taking water from the reactor outside the containment building, cooling it down, and then returning it to the reactor. Because of the damage to the core, the water in the reactor was intensely radioactive, and taking it outside the containment building would have created several additional opportunities for leakage from the plant. The depressurization might have served a useful purpose in venting some of the hydrogen gas that had accumulated inside the reactor as a byproduct of the Zircaloy-water chemical reactions that had occurred earlier, since the gas was one of the factors interfering with the normal flow of cooling water inside the reactor. Venting the gas, however, had its own potential repercussions because hydrogen combined with oxygen can form an explosive mixture. In fact, at 1:50 p.m., the buildup of hydrogen gas in the containment building caused an explosion that resulted in a rapid pressure surge in the building. Officials in the control room at the time have varying recollections of the event. Some say they heard the thud, some ascribed the pressure-increase readings to instrument error, and at least one official thought it might have been a hydrogen explosion. Higgins did not recollect it at all, and the N.R.C. was not informed of the explosion until Friday morning, nearly two days later. The hydrogen explosion itself was a significant indicator of an overheated core, but plant officials, busy with other tasks, ignored the occurrence, and station manager Gary Miller says that he thought the sound he heard might have been from a ventilation damper closing and does not recall being told by anyone about a possible explosion. The explosion, at any rate, did not affect his activities that afternoon. A few minutes after it occurred, Miller left the plant, since he was ordered by his superiors at Metropolitan Edison to go to Harrisburg to deal with the public-relations crisis developing at the state capital.

* * *

Word of the General Emergency had been picked up early that morning by a local reporter who was monitoring the state police radio frequency. The wire services were promptly notified and Metropolitan Edison was soon besieged with inquiries from the press and the public about the nature and seriousness of the accident. Company officials released statements during the day that provided broad assurances that the situation was under control but contained few specific details about the actual circumstances at the plant. Pennsylvania Lieutenant Governor William Scranton, after a telephone conversation with Gary Miller, held a press conference around 11 a.m. in which he repeated the company's advisories concerning the satisfactory conditions at the plant. He was a useful ally for the public-relations effort being executed by Metropolitan Edison officials. Scranton, who had been given no reason to believe that he might be interfering with attempts to bring the accident under control, requested a further briefing from plant officials, and Walter Creitz, the president of Metropolitan Edison, directed Vice President Jack Herbein to go to Harrisburg for this purpose. Herbein, who had flown by helicopter to the TMI-2 site in late morning, had set up an office at the plant's Observation Center on the east bank of the river, just across from the plant. He had been involved in earlier technical consultations with the plant staff—participating in the "senior think tank" sessions via the "squawk box" (telephone speaker-phone) in the TMI-2 shift supervisor's office—but as the day went on, he became the principal company spokesman and devoted much of his time to relations with the news media. Before leaving for Harrisburg, he held his own forty-minute news conference, in which he repeated the company's optimistic assessment of the accident and belittled the possibility of serious damage to the reactor core.

Gary Miller was purportedly unwilling to leave the plant because of the uncertainties about core cooling but, over his objection, he says, he was told to accompany Herbein to the briefing of Lieutenant Governor Scranton in Harrisburg. George Kunder, the TMI-2 superintendent of operations, was also sent. They were gone for over two hours and did not return to the plant until about 4:30 p.m. As they were walking out of the state capitol building after the briefing, they met Herman Dieckamp, the president of General Public Utilities, Metropolitan Edison's parent company, who was there on other business. "My God," Dieckamp exclaimed, "who is watching the store?"

In the meantime, while Miller and the other senior officials were in Harrisburg, the depressurization effort had continued, getting exactly nowhere. The pressure never got low enough so that the shutdown cooling system could be used. In fact, at around two o'clock, when Miller left, the pressure was about as low as it would ever get. Plant officials persisted with the depressurization, despite warnings they received from engineers at General Public Utilities, their parent company. These engineers worked for a branch of the company in Mountain Lakes, New Jersey, known as the General Public Utilities Service Corporation, which did routine engineering work to assist in plant construction but had no role in routine plant operation and no assigned responsibilities during emergencies. Robert Arnold, the vice president in charge of this company engineering group, was informed of the accident around 8 a.m., and engineers working with him received reports throughout the day. By early afternoon, Arnold's staff had recognized superheated steam conditions and had concluded that the reactor appeared to have some type of steam bubble inside it such that the core might be uncovered and inadequately cooled. Shortly after 2 p.m., Arnold called the TMI-2 control room to express these concerns. He spoke to Lee Rogers, the Babcock & Wilcox site representative who was functioning as a member of Miller's "senior think tank." Arnold later said that he had expressed the concern of his staff that "the temperatures indicated that the core had been uncovered and that they had steam bubbles in the loops. I remember Lee Rogers specifically responding that he was confident that the core was *not* uncovered and that they had never uncovered the core." Rejecting the analysis from Arnold's engineering staff, officials at the plant persisted with the depressurization maneuver.

Rogers was unaware that his colleagues at Babcock & Wilcox headquarters in Lynchburg, Virginia, had reached the same pessimistic conclusions as Arnold's group. The communications breakdown between the plant and Babcock & Wilcox headquarters—the result, at least in part, of a shortage of telephone lines at the plant—prevented the engineering staff at Babcock & Wilcox, despite what they describe as frantic efforts, from reporting their findings to Rogers or anyone else in the TMI-2 control room. Rogers occasionally was able to make a local telephone call to Greg Schaedel, another Babcock & Wilcox site representative who lived near Three Mile Island. Schaedel remained at home that day so he could relay information from Rogers to the company's Lynchburg headquarters. Around 1:30 p.m., Schaedel phoned Lynchburg to pass along data about the temperature and pressure inside the reactor. The reaction at Lynchburg was immediate. "Everybody . . . was kind of around the table listening to the squawk box," Bert Dunn, the manager of emergency core cooling at Babcock & Wilcox, recalled. The company officials stared at one another as they listened to the data. "That's superheated! The core's uncovered!" Dunn recalls saying, and he adds that "it finally dawned on me that this plant could've been or was perhaps at that time in real trouble." Still, no one could contact the TMI-2 control room, and Schaedel, at home near the plant, waited all afternoon for a call from Rogers. With him in his living room were three Babcock & Wilcox engineers who had been sent by the company in a chartered airplane but were unable to gain admittance to the plant.

Finally, after Miller, Herbein, and Kunder returned from Harrisburg around 4:30 p.m., Arnold discussed with Herbein over the telephone the growing concern of his engineering group about the safety of the reactor. He convinced Herbein that the steam in the reactor posed a serious problem and that the depressurization effort was a failure. He strongly advised abandoning that strategy and trying a more normal method for cooling the reactor. Herbein, who had returned to the Observation Center across the river from the plant, agreed to order the plant staff to raise the pressure in the reactor back up to normal and to try to restart one of the main coolant pumps customarily used to cool the core. He phoned Miller at the TMI-2 control room to convey this message, but Miller resisted the advice because he and the other members of the "think tank" still believed the core was adequately covered. No tape recording of this telephone conversation was made—only calls that went to the Incident Response Center were taped—and the only available description of the major dispute between Herbein and Miller is the rather bland version provided by the two men. After reviewing the testimony provided by each of them, the official N.R.C. report on the accident summarized the argument with the statement that Herbein "imposed" his decision on Miller.

Attempting to restore stable cooling to the reactor involved two steps. The first was easy: all the operators had to do to restore normal pressure in the reactor was to close the valves they had been using to drain the reactor. The reactor, thus bottled up, would quickly increase in pressure because the remaining water inside it was still being heated by the decay heat produced by the core. The repressurization began around 5:30 p.m., and the pressure in the reactor was back to normal within an hour. The difficult part of this strategy, however, was the second step: restarting one of the four main reactor-coolant pumps, which had failed early in the accident because of all the steam in the reactor. Officials were worried that they would not be able to get one of them going again. By about 6:30 p.m., when telephone contact with Babcock & Wilcox headquarters was finally established, plant officials were able to consult with the engineers there about the best way of getting one of the pumps into service.

The main reactor coolant pumps at TMI-2 are nine-thousand-horsepower, electrically powered, single-speed centrifugal pumps. They are complicated devices and will work properly only if they are adequately cooled and lubricated and if the pressure of the reactor-cooling water is high enough so that steam bubbles do not form inside the pumps. While these pumps serve the centrally important function of circulating cooling water through the reactor during normal operation, they are not supposed to perform any critical role during accident conditions. They were designated as "nonsafety equipment" when TMI-2 was licensed, a decision made by plant designers to avoid the extra cost of meeting the elaborate high-reliability requirements imposed on "safety-related" equipment. The designers, who did not opt for high reliability, got what they paid for: a pump that the TMI-2 operators were unable to start because some of its ancillary equipment—an oil lift pump that needed to be started before the pump itself could be started—was not working. There was an extra oil pump, but it was located in the TMI-2 auxiliary building, which had become a high-radiation zone because it was partly flooded with radioactive water from the reactor. Technicians, nevertheless, were sent into the building to put the additional oil lift pump into service so that a main reactor-coolant pump could be turned on.

By 7:33 p.m., the preparations had been completed for an attempt to start reactor-coolant pump 1A. TMI-2 officials were now in continuous telephone discussions with Babcock & Wilcox headquarters, where engineers had been helping to devise a plan for starting the pump. There was concern about the possible immediate effect that starting the pump might have, so they decided to turn it on for just a few seconds to see what happened. The operators "bumped" the pump and ran it long enough to see an immediate improvement in the cooling of the reactor: a sharp drop in the reactor temperature readings. Satisfied with the test, at 7:50 p.m., nearly sixteen hours after the accident began, they turned on main coolant pump 1A and left it on. To the very great relief of the TMI-2 staff, the TMI-2 reactor was now benefiting from a stable method of cooling.

The successful maneuver was reported to Babcock & Wilcox headquarters, to Arnold's engineering group at General Public Utilities, and to the N.R.C. By 9 p.m., Arnold's staff and the Babcock & Wilcox team had all gone home, and the remaining personnel in the Incident Response Center were all breathing a little easier. "We are fairly relaxed now about the reactor itself," Lee Gossick, the head staff officer at the Incident Response Center, noted in a telephone conversation later that night with Commissioner Richard Kennedy. Even Victor Stello, the most anxious member of the response-center staff, was satisfied that "the biggest part of the problem is behind us."

* * *

The accident was reviewed the next day, Thursday, March 29, 1979, at a formal meeting of the Nuclear Regulatory Commission. Chairman Joseph Hendrie and the four other members of the commission were given a fifty-five-minute briefing by senior members of the agency staff, beginning at 9:55 a.m., after which the chairman and the team of briefers went up to Capitol Hill to give Congress a special noontime report on the accident. The mood of the commission and the staff was hardly euphoric, since the occurrence of such a serious accident would undoubtedly blemish the agency's reputation and add to the already substantial controversy over nuclear-plant safety, but the general N.R.C. assessment of conditions at the plant was definitely upbeat. The crisis was thought to be over with a satisfactory "recovery" operation in progress. The staff told the commission that the "off-site airborne radioactivity" stemming from the accident "has resulted in minimal exposure to the public" and that all measurements "indicate that there is no immediate threat to health and safety." The plant operators were said to be "progressing" in controlling the contaminated water in the auxiliary building at Three Mile Island, which was believed to be the source of the radiation releases that had taken place, and it was thought that the radiation release rate would be "dying off over a period of a couple of days." Other than sending additional officials up to Three Mile Island to "look at the recovery operation," no further N.R.C. action was thought to be necessary at that point. The "primary lead" for directing activities at the plant was still assumed by the Metropolitan Edison Company, the commissioners were told, and the company "had yet to decide on the course of action" it wanted to follow. The staff assured the commission that the N.R.C. would "have a chance to comment" on the plan being developed by Metropolitan Edison officials. The staff also announced that it was assembling a "special investigating team" to examine the event.

As for what exactly had happened at Three Mile Island

the previous day, the staff report to the commission offered few details. "We know no details on the actual events at the plant—the extent, the sequence, the timing, the degree to which we lost the main feedwater, auxiliary feedwater, and, in fact, the sequence of events, as I am going through them, as I said, are really pieced together, based on information that we had before us, literally from yesterday morning until this morning. . . . It is very preliminary," Darrell Eisenhut, the deputy director of the N.R.C. Division of Operating Reactors, told the commission. Between 4 a.m., when the accident started, and around 7 a.m., when leaking radiation was detected, Eisenhut reported, "There is now a gap in the sequence of events that we have."

The commissioners for the most part listened passively to the staff, which is their customary practice given their own lack of technical training, and did not puzzle over the significant aspects of the accident. Moreover, none of them asked about the superheated steam conditions that Victor Stello had noted the day before as evidence that the core may have been uncovered during the accident. The commission knew of Stello's concerns because some of its members had spent much of the day in the Incident Response Center and had personally listened to him describing his concern about the possibility of a badly damaged core. Stello himself did not attend the commission briefing; after spending all day Wednesday and then staying the night at the Incident Response Center, he had gone home to bed.

An hour after the N.R.C. staff briefed the commission, Chairman Hendrie, accompanied by members of the N.R.C. staff, briefed members of Congress who wanted the details of the accident. Hendrie appeared before the Subcommittee on Energy and the Environment of the House Interior Committee, which is chaired by Congressman Morris Udall of Arizona. Television network camera crews, reporters from major national and international newspapers, and an overflow general audience packed Room 2141 of the Rayburn House Office Building. Congressman Udall made a short introductory statement, characterizing the accident as "another in a series of events that lends credence to the contentions of those who think that we have rushed headlong into a dangerous technology without sufficient understanding of the pitfalls," and opened the session to what stretched out into almost two and a half hours of questions and answers. Chairman Hendrie and his senior aides, facing sharp and sometimes hostile questioning from the assembled congressmen, offered assurances that the plant was in a safe and stable condition and that public safety had never been in jeopardy during the accident. Hendrie and the senior N.R.C. staff officials—the same officials who had briefed the commission that morning—went far beyond the cautious and restrained optimism of their morning meeting; indeed, the N.R.C. staff testimony before Congress on several key points sharply contradicts that morning's N.R.C. staff report to the commission as well as some of the information transmitted from Three Mile Island to the Incident Response Center the previous day.

Congressman Don Clausen of California asked Darrell Eisenhut of the N.R.C. staff, "Mr. Eisenhut, I would like to draw on your background and ask you a question that I think is on the minds of many Americans. Are you satisfied that the situation in Pennsylvania is under control and that there is no danger of a core meltdown?" Mr. Eisenhut replied, "Yes, sir, based on the information we had available. The latest we heard was about 10 a.m. this morning. I have every reason to believe that the plant is in a stable condition and that the systems are performing in a way that we think is well understood at this time." That morning, however, in his briefing of the commission, Eisenhut hardly characterized plant-system performance as well understood. Asked by Commissioner Peter Bradford to identify the "pitfalls" of getting Three Mile Island onto the shutdown cooling system, Eisenhut replied, "We are not really sure. There are a number of system problems that seem to have arisen as time goes on.... There has been enough anomalous behavior that we don't want to speculate too much." As for the "stable condition" that Eisenhut assurred Congressman Clausen prevailed at TMI-2, all Eisenhut would say that morning was that conditions were "somewhat stabilized" when the reactor-coolant pumps were restarted and that "a somewhat more normal cooldown mode" had been achieved. Yet, in his answers to congressional inquiries, Eisenhut asserted, "The plant has been followed by a system standpoint, as it is being shut down very normally. At this point, we think the plant is in a *very stable* configuration and we think, from a system standpoint, we understand very well what is going on [emphasis added]."

The conflict between what the N.R.C. knew and what N.R.C. officials told Congress was particularly evident in the basic description of the accident given by Eisenhut during that briefing. Stressing the allegedly "very normal" procedures used to shut down and cool the reactor, he told the congressmen, "The situation from yesterday morning [March 28, 1979] through yesterday and through the night to this point is that the licensee has been in a deliberate slow process of cooling the plant down." He made no mention of the vacillating core-cooling strategy formulated by the Three Mile Island operators. He simply commented that the N.R.C. hoped to see the plant on the shutdown cooling system "in the upcoming hours."

N.R.C. Chairman Joseph Hendrie gave the congressmen terse, emphatic answers, assuring them repeatedly that the plant was in a safe condition. He dismissed the high radiation reading from the monitor in the dome of the containment chamber as an "oddball" instrument error. He denied that there was any serious "ongoing problem" at Three Mile Island, saying that the "trace" releases of radioactive gases would come down and that the plant was not in "a situation of rising difficulty." As to the level of danger the previous day, he said, "I do not think we have been anywhere near a meltdown in this incident." He admitted that there had been some "cracks" in the fuel rods, causing a radioactive leak into the containment chamber surrounding the reactor, but said that there had been no extensive damage to the core.

Hendrie's broad claims provoked skeptical responses from some of the congressmen. Congressman James Weaver of Oregon, an outspoken critic of nuclear power, challenged Hendrie: "You do not really know what is happening, do you?" Hendrie replied that "necessarily somewhat of what we offer here is speculation," but he repeated his prior testimony that "we were nowhere near a core meltdown." Congressman Edward Markey of Massachusetts pressed Hendrie to say how far the water level in the reactor had fallen. The laconic N.R.C. chairman replied, "Don't know." "If you do not know," Congressman Weaver interjected, "then how can you say whether or not we were close to a core meltdown? If you do not know that information, then how can you say that?" Hendrie replied, "Because, I guess, based on the knowledge and instincts of a career in nuclear engineering."

Of all the assurances given the congressmen that day, the most remarkable was Dr. Hendrie's summary judgment on the extent of damage to the Three Mile Island reactor core. He told the congressmen that the "most likely situation in view of the limited information we have on the nature of the fission product releases" into the containment building was that "perhaps about one percent" of the fuel rods "might have cracks." What specific technical basis Dr. Hendrie had for this "one percent" fuel-damage estimate remains unclear. No such estimate was reported in the briefing the commissioners received that morning from their staff technical experts. Indeed, when the staff experts were asked to make just such a summary judgment, they had declined to do so. Commissioner Ahearne had asked that morning, "What can you say about the core? Anything yet?" and Eisenhut had replied, "We can't really say too much about the core, except we can make one inference from the [radioactivity]. The activity levels that we have seen inside the containment would [imply] that we have had fuel failure. To the degree of fuel failure, it's just unclear." Why the situation was any clearer to Dr. Hendrie has not been resolved, as he has declined repeated requests for an interview on this and other subjects related to the accident and the commission's handling of it. His estimate, apparently, was just a general feeling based on his "knowledge and instincts" gained during his career in nuclear engineering, one such instinct possibly being the desire to calm public concerns about the accident.

Near the end of the March 29 special hearing on the Three Mile Island situation, Congressman Clausen stated:

> I want to note for the record that I believe you have been most responsive to our questions and have helped to alleviate the concerns that a lot of us have. I think you have credited yourselves very well in the eyes of the committee

in the presentation of the facts. . . . If the core keeps its cool as well as you gentlemen have kept your cool in your presentation of the facts, it will be all right.

* * *

Senior Metropolitan Edison officials were similarly occupied the day after the accident by demands that they explain what had happened at the plant. They held a major press conference, briefed government officials, and made national television appearances, trying at every opportunity to convey an optimistic view of the status of the plant. Company spokesman Jack Herbein, less familiar than Chairman Hendrie with the finer points of public relations and inexperienced in dealing with the press, had great difficulties as he was questioned by more than one hundred reporters at the company's Thursday morning press conference. (The movie *The China Syndrome,* which had been released only a few weeks before the accident, probably added to his difficulties; it features Jane Fonda as an aggressive reporter who challenges a utility company that is "covering up" safety problems at a nuclear plant.) Herbein, as his own ninety-minute press conference wore on, finally responded to insistent questions about the severity of the accident with the blunt, angry declaration: "I can tell you that we didn't injure anybody in this accident, we didn't overexpose anybody, and we certainly didn't kill a single soul. . . ."

Company President Walter Creitz, somewhat more calmly, conveyed the same message that morning on N.B.C.'s *Today Show* and A.B.C.'s *Good Morning America.* I had also been asked to appear on the two programs, as a representative of the Union of Concerned Scientists, to comment on the accident. Just prior to broadcast time, I talked by telephone with the Incident Response Center but had obtained very little specific information about conditions at the plant, so I took the opportunity during each of the two programs to pursue Creitz for details about the accident. I asked him to identify the equipment in the plant that had been disabled by the accident, and he said that he did not have that information. I asked which equipment was still working, and he didn't know that either. If he didn't know what equipment was working and what was not, I asked, how did he know that the plant was under control? Creitz replied simply by repeating his earlier assertion that the plant was under control. I did not want to make any statement on the air that would needlessly alarm people living near the plant, but the nonspecific answers from both Creitz and the Incident Response Center personnel led me to worry that the plant might not actually be under control. I expressed my doubts that the accident was really over, but said the N.R.C. might have information that showed it was, and if they did, they should make it public immediately. Further detailed information on the situation was not released by the N.R.C. that day.

As the agency's workday drew to a close that Thursday, the only major item of TMI-2 business concerned a minor problem. Plant officials wanted to resume the dumping of waste water—from toilets, showers, and drains—into the Susquehanna River. They customarily dispose of water in this fashion and are allowed to do so because the amount of radioactive materials contained in the water is within allowable N.R.C. disposal limits. The dumping had been suspended on Wednesday morning, however, during the General Emergency, and Metropolitan Edison applied to resume routine dumping on Thursday afternoon. State officials and N.R.C. officials at the plant gave permission to the company to do so, but Chairman Hendrie, when informed of the dumping later in the afternoon, worried about the adverse public-relations impact of the procedure. In a telephone conversation with Edson Case at the Incident Response Center, Hendrie noted his concern, and Case agreed that the dumping was "bad P.R." Officials at the Incident Response Center issued a directive to halt the dumping.

* * *

On Thursday afternoon, about the time Chairman Hendrie was talking with Edson Case about the minor public-relations problem, technicians at TMI-2 were gathering technical evidence that would ultimately be of far greater embarrassment to the N.R.C. chairman. The data acquired by the technicians would completely undermine Hendrie's reassuring noontime testimony before Congress about "one percent" fuel damage and would also negate all the optimistic official statements about the stability of the reactor. The work undertaken by the technicians at Three Mile Island involved an attempt to draw a small sample of the water inside the TMI-2 reactor. Officials had wanted the sample since the previous day, but had been unable to get it because of possible radiation exposure of the technicians who had to obtain it. (Automated sampling equipment is not required by N.R.C. regulations and none had been provided by plant designers.) At 4:15 p.m. on that Thursday afternoon, the plant's radiation-protection foreman and chemistry foreman, both of whom received hefty radiation doses in the process, succeeded in obtaining a one-hundred milliliter sample—a little less than three and a third ounces. It showed that the reactor's normally highly purified water was now laced with massive quantities of radioactive chemicals. This lethal brew contained many of the fission products—radioactive wastes produced by nuclear chain reactions—that normally reside inside the uranium-fuel pellets encased in Zircaloy tubes. N.R.C. officials at the plant immediately concluded that evening, when they saw the results of the reactor coolant sample, that there had been "gross damage" to the reactor's uranium-fueled core resulting in a large release of fission products into the cooling water. They then knew, as Norman Moseley of the N.R.C. later said, "that it had to have been that the core was uncovered partly early [on Wednesday] in order to get the core damage." There is no other mechanism that one could postulate that would cause such a fission-product release.

The definitive evidence of core damage provided by the primary coolant sample helped N.R.C. officials in their reexamination, from late Thursday night through the early hours of Friday morning, of the other data that had been attributed to instrument error or otherwise discounted in their initial assessments of the seriousness of the accident. They focused, in particular, on the temperature data provided by the thermocouples installed in key locations inside the reactor coolant system. "It's too little information too late, unfortunately," Dr. Roger Mattson, one of those analyzing these data, told the commissioners on Friday, "and it is the same way every partial core meltdown has gone. People haven't believed the instrumentation as they went along. It took us until midnight last night to convince anybody that those goddamn temperature measurements meant something. By four this morning, B & W [Babcock & Wilcox] agreed." The temperature readings, in fact, showed that superheated-steam conditions had prevailed for much of Wednesday, a situation that could only arise if the core was uncovered and overheating badly. Stello had been right after all, Mattson and other N.R.C. analysts discovered.

As interpreted in the early hours of Friday morning by Mattson, who was tired after his night in the response center and apparently somewhat unnerved by his findings, the situation inside the TMI-2 reactor core involved "a failure mode that had never been studied" and that was "just unbelievable." He told the commission in a noontime phone conversation on Friday, "My best guess is that the core uncovered, stayed uncovered for a long period of time. We saw failure modes, the likes of which have never been analyzed. It isn't like a LOCA [loss-of-coolant accident]. [It's] some kind of swelling, rupture, oxidation near the top of the quarter center of the [fuel] assembly." Commissioner Gilinsky asked Mattson how long the core might have been uncovered during the first day of the accident.

> MATTSON: Now, unfortunately, the front-end information for us here is very sketchy and it is impossible for us to go back and get it, because you can't distract those people from what they are doing now. So we are guessing on the first four hours, roughly. But, if we are guessing right, it may have been uncovered for as long as fifteen hours.
>
> GILINSKY: And presumably if you had all this metal-water reaction, you must have been up to something like two thousand degrees Fahrenheit—is that right?
>
> MATTSON: We are estimating now that we probably melted some fuel. We are estimating. I don't know what I would say if a reporter called me and asked me if we melted some fuel. . . . It is a severely damaged core.

Having a core severely damaged by the destructive chemical reactions that took place when the fuel rods overheated meant that hydrogen gas, an inevitable byproduct of these reactions, was present in the reactor. N.R.C. analysts were familiar with the basic chemistry of these reactions since this subject had long been a part of general discussions of the phenomena that are expected to take place during serious loss-of-coolant accidents. According to the accepted scenario for such accidents, however, the reactions occur only for a few minutes, even under the worst "hypothetical" conditions, and are terminated so quickly that no large quantity of hydrogen is produced. (Emergency cooling systems, in other words, are supposed to douse the core with water and stop the overheating of the fuel long before any worrisome amount of hydrogen is created.) Mattson and his colleagues, in reviewing data from TMI-2, suddenly found themselves with a new situation: a loss-of-coolant accident that did not take place according to the script. The overheating of the core was not a brief episode of a few minutes, but a protracted crisis that had gone on for hours, producing a correspondingly large quantity of hydrogen. They calculated that, as of late Thursday night, there was about one thousand cubic feet of hydrogen gas in the reactor. The gas would be expected to collect in the highest parts of the reactor, in particular in the "upper head" of the reactor vessel above the core. The upper head of the TMI-2 reactor has a volume of only slightly greater than eleven hundred cubic feet, so the hydrogen gas bubble, according to N.R.C. estimates, was almost big enough to fill the top of the reactor completely. None of the official safety analyses that had ever been performed for pressurized-water reactors had analyzed the hazards posed by such a bubble. The hasty analysis carried out by Mattson and his fellow N.R.C. safety experts late Thursday night and early Friday morning led them to worry that when the reactor pressure was lowered so the reactor could be put on the shutdown cooling system, the hydrogen gas bubble would expand proportionately. The expanding gas bubble could blanket the core and lead once again to inadequate core cooling. Mattson told the commission that the N.R.C. had "every systems engineer we can find" thinking about "how the hell" to get the bubble of gas out of the reactor. "Eventually the bubble is going to grow," Mattson said, leaving the commission in a "horse race" between how fast it grew and how quickly they could find a way of getting the gas out of the reactor. The question, he told the commission, was "Do we win the horse race or do we lose the horse race?"

* * *

By Friday morning, March 30, the general impression that the TMI-2 accident was "over" had completely disappeared. The frantic technical reassessment of the situation, which N.R.C. officials had begun late Thursday night, was only one aspect of the renewed General Emergency. At the plant itself, another set of difficulties—to be followed by another set of mistakes by the N.R.C.—also brought the threat of a new crisis, one that would lead, in fact, to near-panic in the surrounding area. The problem at the plant centered on the vast amount of radioactive debris created by the accident and on the troubles the operators were having in keeping this hazardous material safely confined within the plant. The damage to the core early in the accident had dumped huge quantities of radioactive materials—about ten percent of the total inventory in the core—into the reactor's cooling water. Some of this material escaped into the containment building housing the reactor, the bottom of which was now flooded to a depth of several feet with radioactive water, and some of the water was also inadvertently pumped into tanks in the plant's auxiliary building, where it had overflowed. (This happened because automated plant systems had pumped thousands of gallons of radioactive water from the reactor into a receiving tank that was already nearly full.) Radioactive gases were also released from the damaged fuel, and they posed the difficult storage problems that the operators had to deal with in the predawn hours of Friday morning. Radioactive gas from the reactor was collecting in one of the reactor's auxiliary water-supply tanks, known as the makeup tank, and it was threatening to overpressure this tank and force the operators to rely on an alternative tank. They did not want to do this, however, since the alternate water supply was also the source for the emergency cooling water that might be needed if subsequent equipment malfunctions led to any problem in cooling the core. The operators, accordingly, were trying to transfer radioactive gas from the makeup tank, where it didn't belong, into a tank specifically intended for the storage of radioactive gas. They were experiencing difficulties because the connecting piping going to the gas-storage tank was leaking. Every time they tried to transfer the gas, a small "burp" of radioactive gas would escape into the plant's auxiliary building, and the ventilation system in the building would then discharge the gas into the atmosphere. They finally decided that the buildup of radioactive gas in the makeup tank was getting out of hand and that they would have to vent the radioactive gas from it quickly, into the gas-storage tank. Instead of venting small amounts at a time in repeated transfer attempts, as they had been doing, they decided that they must open the vent valve and leave it open—thereby permitting an uncontrolled radiation release from the plant due to leaks in the transfer pipes. Plant officials called for a helicopter to monitor the unavoidable radiation release, and they notified state health officials and the Pennsylvania Emergency Management Agency, which is responsible for civil defense, that there was a "planned but uncontrolled release" of radioactive gas from the plant. The situation in the control room at this time was confused, and officials there had conflicting impressions, for example, about who was managing the operation. James Floyd, the operations supervisor at TMI-2, thought he was in command, but other officials thought they were giving the orders. Both Floyd and another official, for example, made separate telephone calls for the helicopter, and they later gave separate—and conflicting—advisories to state authorities.

Around 8 a.m., an hour or so after the venting began, the TMI-2 control room received a report from the helicopter, which had just taken a reading of the radiation level one hundred and thirty feet above the auxiliary building's exhaust stack. The reading in that location indicated twelve hundred millirems per hour. This was not actually very different from other readings that had been taken the day before while earlier attempts to vent radioactive gas were in progress, and by the time the wind had blown the gas off the site and diluted it, the resulting doses to the people living in the surrounding area would be inconsequential. Floyd, however, became very excited and picked up the telephone to alert the Pennsylvania Emergency Management Agency to the situation and to advise them of the possibility that an evacuation of people downwind of the plant might be necessary. He failed to reach officials there and placed another call, to the Dauphin County Civil Defense Agency, and asked that they get a message to the other agency. Floyd finally talked with the staff of the Pennsylvania Emergency Management Agency around 8:40 a.m., and officials in the agency, who later described Floyd as overwrought and extremely excited, quickly passed the information he had provided on to other state officials. Shortly after 9 a.m., Governor Thornburgh's staff asked Karl Abraham, an N.R.C. public-relations man who happened to be in the governor's offices, to call the N.R.C. to verify the reports of a twelve-hundred-millirem-per-hour radiation level at the plant. Abraham's call to the Incident Response Center apparently caused Denton and other senior officials there to panic. What Abraham told them was that there had been a report of "an uncontrolled release of airborne [radio] activity from a release point in one of the cooling towers" and he asked them to verify this. Yet, instead of interpreting it as a request for information, officials took Abraham's inquiry as a report of a twelve-hundred-millirem radiation level at the plant. They paid no attention to the fact that the "cooling towers" themselves do not release radioactive material, so the question Abraham posed was technically nonsensical. Instead of having the Incident Response Center staff call the plant to check up on the report, the senior N.R.C. officials, led by Denton, who has said that he "sort of assumed that [the report] had been verified," held a hasty discussion on whether to order an immediate evacuation.

The confusion among senior N.R.C. officials was further compounded by a garbled report relayed from Inspector Higgins at the plant shortly before Abraham's call. According to the report, TMI-2 operators were dumping the contents of the radioactive-waste storage tank, "which is causing a considerable release rate," and "civil defense" and state health authorities were being notified. There was some type of mixup connected with this message because there was no venting of the radioactive-gas storage tank itself; what was happening was the venting of the gas in the makeup tank *into* the storage tank, resulting in incidental radiation releases because of the leaks in the transfer pipes. N.R.C. officials were led to believe that the entire inventory of stored radioactive gases was being deliberately released, rather than the relatively small amount from leaking transfer pipes. This "report" from Higgins, which was written down on a message form, had been relayed from the Region I N.R.C. office (near Philadelphia) and was not verified by Denton or other senior officials.

Six minutes after Abraham's call, Denton acted. At 9:15 a.m., after reviewing the matter "abruptly and with little or no deliberation" (according to a later N.R.C. report), Denton turned to Harold Collins of the N.R.C. Office of State Programs and told him to advise state officials that the N.R.C. recommended an immediate evacuation of the area around TMI-2. Denton did so casually, as if he were merely sending an aide out to get him a cup of coffee. The other two ranking members of the Incident Response Center's "Executive Management Team," the nominal N.R.C. emergency decision-making group, were unaware of Denton's order to Collins. One of these officials was on the phone, trying to get through to the commission to discuss a possible evacuation recommendation, and the other was not present at the moment Denton made his snap judgment. Denton did not give Collins any specific instructions on how large a region needed to be evacuated. He just left this up to Collins.

The phone call from Collins to the Pennsylvania Emergency Management Agency, which was headed by Oran Henderson, who had been an infantry colonel in Vietnam, touched off a rapid series of events that left many people living near TMI-2 badly frightened. Henderson notified the lieutenant governor (who then briefed the governor) and also the Pennsylvania Bureau of Radiation Protection, which had been monitoring the plant. (Officials in the bureau insisted that they had no information to justify an evacuation, but would check conditions at the plant once again.) Henderson also called Kevin Molloy, the Dauphin County Civil Defense chief, who promptly alerted county agencies, as well as the local radio station, WHP, selected as the primary emergency broadcast station for the area. The station then carried an announcement that an evacuation order for the area might be imminent. Many people did not wait for the formal order but rushed to leave the area following the broadcast.

Around 10 a.m., TMI-2 operator Ed Frederick, who was then off duty, drove into Middletown to buy sandwiches for the TMI-2 staff, and he could not understand why people were rushing about so hurriedly. At the sandwich shop, the proprietor, who was just about to close up shop and flee, told him of the radio announcement. Frederick called the TMI-2 control room from a pay phone and relayed the news. Gregory Hitz, who answered the phone, exclaimed, "You've got to be kidding!" TMI-2 personnel immediately confronted the N.R.C. officials at the plant and heatedly asked them what on earth the N.R.C. was doing since there was nothing going on to justify an evacuation. (Plant officials were apparently unaware that TMI-2 operations supervisor Floyd had also played a role in triggering the evacuation directives.) Inspector Higgins told station manager Gary Miller that he had no idea what was going on, and quickly consulted some of the N.R.C. personnel at the site, who also knew nothing about the evacuation recommendation. Inspector Charles Gallina, the radiation specialist at the site, could not understand it at all because, from what he could see at the plant, he has explained, "preventive actions were being taken" to prevent further "puff releases" of radioactive gas. He thought that the evacuation recommendation was a "total mistake."

The Pennsylvania Bureau of Radiation Protection, within a half hour or so of Collins's phone call to Henderson, had checked out the report of a major release of radioactive materials from the plant, and found that it was spurious. The bureau could not get a telephone call through to the governor's office or to Henderson's office, however, to report its findings: the line was busy because of all the telephone calls that followed the radio announcement. Bureau of Radiation Protection officials, however, did manage to contact Collins at the Incident Response Center and protested the evacuation recommendation. They had already talked with Gallina at the site, they said, and he concurred that there was no basis for the recommendation. The state officials asked Collins to "relay the information" to senior N.R.C. officials that "you screwed our situation up . . . incredibly." Collins ruefully explained that the N.R.C. had "a lot of big wheels sitting here around tables" and that he was low on the "totem pole" and was only following orders when he conveyed the evacuation directive. "Yeah, well anyway," he added, "I understand the chairman is calling the governor. And what the chairman's going to say to the governor I have no idea." The final report of the special N.R.C. study group that investigated the TMI-2 accident summarized Denton's 9:15 evacuation order under the heading "The Friday Morning Fiasco."

* * *

Shortly after dispatching Collins to issue an official recommendation for evacuation on behalf of the N.R.C., Denton talked by telephone with the N.R.C. commissioners. The five commissioners had assembled, as if it were a normal workday, at their offices in downtown Washington. After a quick briefing on what the N.R.C. staff characterized as a deteriorating situation at TMI-2, the commissioners decided to go into closed emergency session to consider how they would respond to the new state of crisis. The public was barred from these commission meetings, which continued for the next several days, but the "Government in the Sunshine Act" required that records of all meetings be maintained. Tape recordings were therefore made of the sessions, from which official transcripts were later prepared. The transcripts are not a complete record of the commission's deliberations because many conversations between the members—in the hallways, the elevators, or the men's room, for example—were not taped. The tapes, nevertheless, like those from the Nixon White House that figured so prominently in the impeachment proceedings and Watergate trials, provide an important first-hand account of the commission's handling of the accident.

Denton told the commissioners that "[We] did advise the state police to evacuate out to five miles." (Actually, this is a rather inaccurate and off-hand report of what had been done. It was not the state police but the Pennsylvania Emergency Management Agency to whom the recommendation had been given, and Collins had recommended a much broader evacuation—out to ten miles—from the plant.) Governor Thornburgh, however, wanted to hear directly from the chairman of the N.R.C. whether the N.R.C. was recommending such an evacuation, so the commission deliberated on what Chairman Hendrie should say to Thornburgh. The commissioners were far removed from the situation, had scanty information on which to base their judgment and little technical competence of their own, and were hampered by poor communications. They decided, nevertheless, to interject themselves directly into the management of the crisis.

Chairman Hendrie, who had assured Congress the day before that the "intergovernmental agency coordination and liaison has worked well and promptly" and the "communication flow from all parties to all other parties" was "in good shape," now told Denton, "I don't know what you can do to improve the communication situation, but it is certainly lousy." Denton complained that "we don't know what they are doing" at the plant. "They may be taking just the proper action," he said, but on the other hand, he didn't have any basis for believing that another uncontrolled radiation release would not happen again. The N.R.C. was getting information "second- and third-hand," he said, and kept trying to "second-guess" the company on what should be done. "It is really difficult to get the data," he told the commissioners later that morning:

> We seem to get it after the fact. They opened the valves this morning, or the let-down, and were releasing [radioactive materials] at a six curie per second rate before anyone knew about it. By the time we got fully up to speed, apparently they had stopped, there was a possible release on the order of an hour and a half.

Chairman Hendrie asked Denton whether Richard Vollmer, who had gone to Three Mile Island on Thursday to be the N.R.C.'s senior representative at the site, was "on top of" the situation, and Denton replied, "Well, I sure hope so, but he is not in the dialing communications line and I have not been able to reach him." Denton said that N.R.C. officials sent up to the site seemed, because of the poor communications, to "fall into a morass—it seems like they are never heard from."

Even the most basic data needed to assess the possible threat to the surrounding population from releases of radioactive material from the plant, such as which way the wind was blowing, were unavailable to the commissioners. "What is the wind speed, do you have any idea?" Commissioner Gilinsky asked Denton.

DENTON: We are trying to establish that.

GILINSKY: And when did this plume—when was the puff released?

DENTON: Within the last two hours.

CHAIRMAN HENDRIE: Presumably it has just terminated recently then.

DENTON: We don't know how long, but if it was a continuous release over a period of an hour or an hour and a half, which—from what I understand—which is kind of a lot of puff.

GILINSKY: So even with a modest wind . . .

HENDRIE: A couple of knot wind and the damned thing—the lead edge is already past the five mile line. . . .

Denton wanted the commission to take action so that people could be moved out of their homes before the spreading plume of radioactive gases passed overhead. "I think the important thing is for evacuation to get ahead of the plume—is to get a start rather than sitting here waiting to decide," he told the commissioners. (The official, but unedited, transcript of this conversation says "waiting to die," but Denton believes he said "waiting to decide.") Chairman Hendrie was concerned that, because it wasn't known where the winds were carrying the released radioactive materials, evacuation might move people who had already been exposed back under the plume or might move people from safe to less safe locations. When he was asked by the agency's chief public-relations officer, Joseph Fouchard, "Don't you think as a precautionary measure there should be some evacuation?", he replied, "Probably, but I must say it is operating totally in the blind." About what was the safe direction in which to move people, the chairman said, "I don't have any confidence at all." Fouchard urged that Hendrie call Governor Thornburgh immediately, since the governor's office had called the N.R.C. saying that Thornburgh's information from the plant was "ambiguous" and that he wanted "some recommendations from the N.R.C." "We are operating almost totally in the blind," Hendrie said. "His information is ambiguous, mine is nonexistent and—I don't know, it's like a couple of blind men staggering around making decisions."

With his public-relations man pressing him for a decision, the chairman asked senior staff members in the Incident Response Center for advice. "Is there a consensus there that we ought to recommend to the governor he move people out within the five-mile quadrant?" Denton replied, "I certainly recommended we do it when we first got the word, Commissioner. Since the rains have stopped and the plume is going—I would still recommend a precautionary evacuation in front and under [the plume]." Brian Grimes, an N.R.C. accident consequence analyst, told the chairman that the radiation levels were below evacuation guidelines suggested by the Environmental Protection Agency and that "the most that should be done, in my view, is to tell people to stay inside this morning." Commissioner Bradford, the lawyer on the commission, was the only one of the five commissioners who wanted at that point to recommend a precautionary evacuation.

Not having heard from Hendrie, who was still discussing matters with the Incident Response Center, Governor Thornburgh placed his own call directly to the N.R.C. chairman's office. Hendrie took the call at 10:07 a.m., and the other commissioners gathered around a speaker-phone to listen. The chairman began:

> Governor Thornburgh, glad to get in touch with you at last. I'm here with the commissioners. I must say that the state of our information is not much better than I understand yours is. It appears to us that it would be desirable to suggest that people out in that northeast quadrant within five miles of the plant stay indoors for the next half hour.

Hendrie added that other data coming in soon from the plant would "tell us whether it would be prudent" to begin any evacuation of the neighboring population.

The conversation with Thornburgh was interrupted at this point by a call to the commissioners from the Incident Response Center. The center told the commissioners that it had just learned, among other things, that it had misinterpreted the twelve-hundred-millirem-per-hour reading and that the release of radioactive gas from the plant had stopped. Hendrie, who had just recommended to the governor that people northeast of the plant be told to stay indoors, was informed that the winds were "light and variable" and that they were "headed toward the south," in the opposite direction from what he had assumed.

Getting back to Governor Thornburgh, Hendrie relayed this information and said that "I'm afraid we are behind the event" and that "the emissions from the plant, at least for the moment, have apparently been cut off. I think I would continue to recommend that people stay inside this morning. And as our information improves—hopefully, it will—then we can see where we go from there."

Hendrie did not, however, discuss the case for a precautionary evacuation as a prudent step in light of possible future releases from the plant. (An evacuation in response to some adverse event was obviously much easier to justify. A more general, precautionary evacuation, based on the commission's own uncertainty and lack of information, would have been an embarrassing admission, which Hendrie did not make publicly during the crisis, that the N.R.C. really had very little knowledge about or control over the situation at TMI-2.)

Governor Thornburgh asked Hendrie about the 9:15 N.R.C. call from Collins to state officials. "Was your person, Mr. Collins, in your operations center, justified in ordering an evacuation at 9:15 or recommending that we evacuate at 9:15 a.m., or was that based on misinformation? We really need to know."

Chairman Hendrie replied, "I can't tell what the—I can go back and take a check, Governor, but I can't tell you at the moment. I don't know." He did not tell Governor Thornburgh that it was Denton, the chief N.R.C. staff safety officer, not Collins, who had recommended the evacuation. (Commissioner Bradford, who was listening to the conversation and who had tried to interrupt, said later that he thought Chairman Hendrie had "fudged" his answer to Thornburgh's question.)

Governor Thornburgh pressed Hendrie on the subject of evacuation, asking whether "we have any assurances that there is not going to be any more of these releases?" Hendrie said:

> No, and that's a particularly important aspect I want to talk to you about. As best as I can judge from the kind of information coming through from the plant, it is not clear that they won't get into this kind of situation again [i.e., possible operator actions that might release puffs of radioactive materials]. I trust not again without all of us knowing it in advance and being ready to anticipate what we may need to do.

Thornburgh asked, "But you still think it is not necessary or reasonable to order a precautionary evacuation, just on the event that we have more bursts?" Hendrie thought it "would be just as well to wait" for an announcement by plant officials that "there may be" another release before considering such a step. The conversation ended with Thornburgh once again asking Hendrie to review the basis for the 9:15 a.m. N.R.C. evacuation recommendation. (Thornburgh repeated his question to Hendrie when they met in person two days later, but Hendrie still did not respond and tell the governor the story behind "The Friday Morning Fiasco.")

At 10:25 a.m., after his conversation with Hendrie, Thornburgh spoke on a live radio broadcast over WHP. He tried to calm residents near the plant, whom he advised simply to stay indoors for a while. Hendrie, in the meantime, had been reconsidering, along with the other commissioners, the advice that he had given the governor just a little while before. Edson Case, at the Incident Response Center, had told the commissioners that "the information you will have within the next hour may be as sketchy or less than you had the last time." Case added that "the plant is in a tender state" and the N.R.C. did not really know "what they are doing" at the plant and that he had "no confidence" that the N.R.C. would know in the upcoming hour. The commissioners discussed recommending the evacuation of at least the portion of the population most sensitive to radiation (pregnant women and small children). The general sense of their discussions was that this would be prudent, although there was no clear reasoning that distinguished between the advisability of this selective evacuation and that of a broader precautionary evacuation.

Hendrie called Thornburgh at 11:40 a.m. to report the N.R.C.'s latest round of discussions, and told him, "If my wife was pregnant and I had small children in the area, I would get them out because we don't know what is going to happen." It was decided to make such a recommendation to people living within five miles of the plant.

At 12:30 p.m., having tried earlier that morning to calm public fears aroused by the N.R.C.'s "accidental" evacuation order, Thornburgh held a press conference at which he read, in a grim voice as he looked directly into the television cameras, a new and unsettling announcement:

> Based on advice of the chairman of N.R.C., and in the interests of taking every precaution, I am advising those who may be particularly susceptible to the effects of radiation—that is, pregnant women and pre-school-age children—to leave the area within a five-mile radius of the Three Mile Island facility until further notice. . . .We have also ordered the closing of any schools within this area. I repeat that this and other contingency measures are based on my belief that an excess of caution is best. . . .Current [radiation] readings are no higher than they were yesterday [Thursday]. However, the continued presence of radioactivity in the area and the possibility of further emissions lead me to exercise the utmost of caution.

* * *

As the crisis appeared to deepen, a more elaborate federal response was organized. The N.R.C. Incident Response Center had informed the White House early Friday morning as soon as it learned of the uncontrolled burst of radiation coming out of the supposedly stabilized Three Mile Island plant. Jessica Matthews of the National Security Council staff had prepared a memorandum for Zbigniew Brzezinski, who, in turn, briefed President Carter on the situation. Carter had called both Hendrie and Thornburgh and put the federal government's full resources at their disposal to help control the situation. Hendrie had agreed to send a senior N.R.C. official to be the federal government's representative at the site, and President Carter told Hendrie that the White House Signal Corps would quickly install a special telephone link from the White House and the N.R.C. headquarters to Three Mile Island.

Hendrie had phoned Harold Denton around eleven o'clock on Friday morning and told him that the President would "like to see a senior officer" at the site "and I think you are it." Denton, who said later that he was "somewhat taken aback" and had a gee-whiz-golly-who-me? reaction to the assignment, told Hendrie that he couldn't go alone and would need to take a team of "experts." He did not choose a team on the basis of technical expertise on Babcock & Wilcox plants, he said in a recent interview, but limited his selection to senior N.R.C. management officials from various N.R.C. divisions. (One of Denton's senior aides has said that this decision created "an advice system that was very poor.") Hendrie gave Denton no instructions on whom he should take or what he was supposed to do. "This didn't come up," Denton said later, adding:

> It was more of "We've decided we want somebody really up there who's in authority who can understand and know what's going on and get back to us." And we didn't really ever discuss exactly what my role was going to be, and I didn't want to claim that I was the President's representative, but after the chairman hung up that morning, I did get promptly a call from the White House saying, "Where would you like to leave from?" and "What time?" and "How many people do you want to take?" It's sort of a new experience to get asked those sort of questions.

Denton and twelve of his associates arrived by Army helicopter at the Three Mile Island Observation Center across the river from the plant shortly after 1 p.m. on Friday afternoon.

> We got up there—there was really no place to meet. We got out of the helicopter and we walked over and went over to the Observation Center. Well, the Observation Center was sort of jammed with people and the press were all around the outside. We went in and talked to Jack Herbein just briefly, but he had an office with phones ringing off the hook. He seemed very harassed and took me over to see Bob Arnold, who was one rung up in the management of the G.P.U. organization and used to be the head man at Three Mile Island. He was in a lady's house who just happened to live near the Observation Center and made arrangements to introduce me to the owner of the house. She let me use her living room and gave us all coffee and that was the first time I talked with my people at the site, in her living room. Arnold clued me in as to what he knew at the time, and I then sent the staff off to the plant for everyone to go look in your area and get back to me at six o'clock or some such time. At some time in that period the kitchen phone rang, and it was the President calling.

Denton reported to the President, and back to N.R.C. headquarters, several times on that Friday and throughout the weekend. "This place up here is a madhouse," he told the N.R.C. Incident Response Center on Friday evening, after his staff had reviewed the status of the plant.

> The utility is a little shy, in my view, of technical talent. We outnumber them [with a staff of twenty-two N.R.C. personnel on site at that point]. They are pretty thin, I'm trying to convince them to bring comparable levels from their own organization. Their cooperation is good, but it is obvious that they are a small outfit here and the guys are getting swamped with demand.

In another report, Denton told the commissioners once again that he had "developed a management concern about the capability of the utility here to cope with the new problems that come up. They're stretched very thin in some areas. I've discussed it with the local management and with the management of G.P.U. I think they need stem-to-stern reinforcements down here in many areas." Babcock & Wilcox, Denton said, was mainly just "monitoring" the situation and was not "turned on" to doing the kind of analyses he thought were needed. In particular, Denton indicated that he was trying to "heighten the sensitivity" of the plant management to the necessity for "planning, developing procedures to cope with eventualities rather than waiting for something to fail and then trying to work your way out of it. I would sure like to see them muster their resources ... and tackle these problems clearly. That's kind of where I am today."

The N.R.C. commissioners, who gathered around a speaker phone to listen to Denton's reports, were greatly perturbed by the reported lack of competence at the site and the news that Babcock & Wilcox was not providing the support that Denton said was needed. "I'm flabbergasted both that B & W needs an invitation and that the invitation hasn't been issued," Commissioner Bradford commented. Hendrie placed a call to G.P.U. President Herman Dieckamp to urge a full-scale effort to bring the required people and technical resources to the site to deal with the difficult job of restoring the plant to a stable condition. The chairman waited before calling Dieckamp, however, until presidential assistant Jack Watson "could read him the riot act to get people down there." Dieckamp told Hendrie that he was already arranging to bring in experts from all over the country to form an advisory team on how to cope with the crisis. Experts from the other reactor manufacturers (Westinghouse, Combustion Engineering, and General Electric), and from the Electric Power Research Institute, the various national laboratories, the military, universities, and private consulting firms—impeded, unfortunately, by the United Airlines strike then in progress—began arriving on Saturday and Sunday. They met in the Air National Guard building at the Harrisburg Airport, and continued to meet there because it was felt that it would be better to let them do analytical work free from the commotion at the plant a few miles away.

The industry experts were asked to form themselves into working groups and to focus their attention on a set of urgent questions that Metropolitan Edison needed answered. The initial list of questions given to the group asked, for example, "What is the physical condition of the reactor core with respect to the degree of damage and its coolability?" Another question was "What are the unique problems associated with the cooling system, and in particular, what are the specific problems associated with the [hydrogen] bubble . . . in the reactor vessel . . . ?" Herman Dieckamp recalled giving the Industry Advisory Group the following general instruction at their first meeting:

> Look, I don't know all of you guys in great detail, and I don't know each of your . . . greatest [areas of] knowledge, but I think you yourselves know where you can best contribute . . . . Conglomerate yourselves into these groups that are working on the problems and go to work. That is about as much as I can tell you what to do.

The overall management of TMI-2, despite the influx of N.R.C. personnel and industry experts, was still left in the hands of Metropolitan Edison Company and its parent company, General Public Utilities. The N.R.C. remained largely in its customary "monitoring" role, and the industry advisory group stayed at the airport, feeling, as one of its co-directors, Sol Levy, a nuclear consultant from San Jose, California, explained, "frustrated about our role for the first few days because we were not sure our advice was being received." No clear management of the situation was established by the N.R.C. or the company until about a week after the accident. Edwin Zebrowski, the other co-director of the advisory group, said in an interview that in the first few days after Denton's arrival, the N.R.C. was behaving "like an army in which it was unclear whether the generals, the captains, or the corporals were giving the orders." "I think we've got too many people up there in the control room now," Denton reported back to headquarters at one point. "We just got up a list of how many people were in the actual building. Christ, it's fifteen or twenty people watching their backs." While the N.R.C. was looking over their shoulders, Denton explained, the executives of the parent utility company, General Public Utilities, "were really holding the reins" and were so "fully occupied with day-to-day details" that an overall plan to restore the plant back to a stable condition was not being developed. "I don't mention this to indicate that we ought to take over the operation, because I'm not sure we could do any better, either," Denton added.

* * *

While Denton tried to deal with a management vacuum at the site, N.R.C. headquarters focused on another set of difficulties. One issue that absorbed much of the commission's time on Friday afternoon was public relations. "Well, yesterday I thought we had things in pretty good shape," chairman Hendrie told Secretary of Energy James Schlesinger in a telephone conversation, "and the immediate public-opinion standpoint last night was in good shape. Then this morning there were more releases. . . ." Hendrie was in close touch with White House press secretary Jody Powell, and the commission was also contacted by President Carter's media advisor, Gerald Rafshoon, who was trying to arrange television appearances for the commissioners. Hendrie and Commissioner Kennedy put a great deal of effort into drafting a press release that Kennedy said "has to be reassuring—reassure people that at least we're working on it." They finally worked out what Kennedy called a "first-rate press release," which Commissioner Ahearne observed had a very optimistic "flavor" and was aimed at "counterbalancing" press and television reports about the serious conditions at the plant. ("Which amendment guarantees freedom of the press?" Hendrie quipped at one point. "I'm against it.") The N.R.C. press release acknowledged the severe core damage that had occurred on Wednesday and noted some of the related developments, such as the presence of hydrogen gas in the reactor. There was, the statement emphasized, no imminent danger of a core meltdown. The announcement also noted that Denton and other N.R.C. officials had arrived at the site.

Denton himself had met briefly on Friday afternoon with reporters who had assembled near the site. "Harold does have to chat with these newsmen on the site up here, even if it is in very vague terms, to show the flag, that we're up here," N.R.C. public-relations chief Fouchard told Chairman Hendrie. Fouchard later reported to Hendrie that "Harold just talked very briefly with reporters here because there was no way we could hide him." Denton reported directly to Hendrie, who had been warned by Jody Powell to coordinate statements with Denton so that Denton would not say one thing at the site while "the dumb chairman" was saying another thing in Washington. Denton told Hendrie he had said to the press that "there was no imminent hazard" and that, "Now—they, you know, asked a lot of other questions, most of which I didn't answer."

* * *

"The central question," as Chairman Hendrie called it, that absorbed most of the commission's attention during the weekend following the accident concerned the need for precautionary evacuation. Instead of having to react hastily when conditions deteriorated to the point where they lost control over the reactor completely, the commissioners asked themselves repeatedly whether it might not be prudent just to move people away until the safety of the plant could really be assured. Dr. Mattson made such a recommendation on Friday in a blunt noontime conference with the commissioners. "The latest burst didn't hurt many people," he noted, but said, "I'm not sure why you are not moving people. Got to say it. I have been saying it down here. I don't know what we are protecting at this point." Mattson, one of N.R.C.'s senior safety analysts who holds a doctoral degree in mechanical engineering, explained to the commissioners the situation as he understood it:

> Well, my principal concern is that we have got an accident that we have never been designed to accommodate, and it's, in the best estimate, deteriorating slowly, and the most pessimistic estimate, it is on the threshold of turning bad. And I don't have a reason for not moving people. I don't know what you are protecting by not moving people.

Hendrie conceded that afternoon that "we've got a more serious situation than I thought yesterday by a hell of a long shot," that "the core is in considerably worse shape than we thought," and that "we don't, at the moment, see a way to get the hydrogen out of there neatly." Nevertheless, he continued to resist the notion of recommending a broader evacuation than the advisory given by Governor Thornburgh relating only to pregnant women and pre-school-age children.

After talking with Mattson, Hendrie went to a meeting President Carter had ordered for 1 p.m. in the White House Situation Room, which is only a few blocks away from N.R.C. headquarters. The President had directed that representatives from the N.R.C., all federal agencies with emergency response capabilities, and the White House staff meet to coordinate plans for handling the Three Mile Island crisis. Hendrie briefed the group, which included presidential assistants Watson, Brzezinski, Powell, and Matthews, on the status of the plant—one senior White House aide, in a later interview, dismissed Hendrie's report as "not very helpful"—and the various federal agencies discussed the roles they could play in organizing a satisfactory contingency plan for mass evacuation of the area surrounding the plant. "The federal government performed better than I've ever seen it perform," Matthews said later. "By noon on Saturday, we could have done a very orderly evacuation out to ten or twenty miles from the plant," she said, with provisions made to provide cots, food, and shelter for all the displaced persons.

Federal health agency officials met later on Friday afternoon, at the office of Health, Education, and Welfare Secretary Joseph Califano, and one of the steps they decided to take was to find a drug manufacturer who could produce a massive supply of potassium iodide. This drug, which had been approved "for use in radiation emergency" by the Food and Drug Administration (F.D.A.) in December 1978, would be helpful in preventing thyroid cancers in the event of a large release of radioactive iodine from TMI-2. (The radioactive iodine would concentrate in the thyroid glands of people downwind of the plant; potassium iodide is a "blocking agent" that could help reduce the consequences because it prevents the thyroid from absorbing the radioactive iodine.) Local pharmacies could not meet the needs of residents for the drug, and the F.D.A. arranged with the Mallinckrodt Chemical Corporation to manufacture an emergency supply. The company called in dozens of employees to its plant in Decatur, Illinois, and they worked through Friday night and all day Saturday to manufacture a massive supply of the drug. The company did not have enough small bottles on hand with medicine droppers for bottling the drug, however, and the F.D.A. had to make other arrangements to get these. Air Force cargo jets delivered the first load of potassium iodide to Harrisburg International Airport late on Saturday night. By the following Wednesday, six more shipments had arrived, giving the state health authorities enough potassium iodide to administer ten daily doses to ten million people.

Despite the heroic efforts to supply the drug, the endeavor was not without complications. State health officials found that many of the bottles in the first shipment contained "hairlike" filaments and other debris, possibly from the manufacturing process or from the use of unwashed bottles. (The F.D.A. advised that the drugs, while not up to normal standards, were still fit for use.) In addition, many of the bottles were leaking, unlabeled, provided with medicine droppers that did not fit inside them, and supplied only about half the recommended dosage.

The value of potassium iodide to protect the public from some of the harmful effects of nuclear-plant accidents had been recognized for many years. A major study of nuclear-plant safety done by the American Physical Society in 1975, for example, specifically recommended that the N.R.C. consider taking steps to ensure that enough potassium iodide was manufactured and tested so that it would be available in the event of a serious accident. The N.R.C. had ignored this recommendation, which is why such a hectic last-minute effort to produce it in sufficient quantities had to be undertaken during the TMI-2 accident.

By Saturday, the White House believed that the logistics for a mass evacuation had been satisfactorily arranged, although there was some skepticism at the N.R.C. about the ease of moving so many people quickly enough "if the balloon went up." The big question remained whether, and on what basis, such an evacuation should be ordered. "It would be easy if things turned sour to order an evacuation—in other words, if the situation became acute," Matthews said. "The really difficult problem that we struggled with all through the weekend was whether a *precautionary* evacuation should be ordered." Information from the N.R.C. kept changing, and White House sentiments for and against a precautionary evacuation shifted accordingly. "The general attitude that prevailed," Matthews said, was partly one of "Today it's a bubble, tomorrow it could be God-knows-what." "On the other hand," she added, "people at the site and at the upper levels of N.R.C. felt improvements to the situation were being made," and the reports of these improvements lessened the apparent urgency of an evacuation. There were costs to an evacuation, she noted, but said that the feeling at the White House was that the costs would not be too high "if there were some finite ending." But if the federal government, merely as a general precaution, advised people to leave, when could they be told it was safe to return? There were also, inevitably, political considerations pertaining to the President's energy program, which stressed increased nuclear-power-plant construction. "One can't eliminate" the political aspect, Matthews said, but stressed that this was "not an overriding concern. We felt that we were under intense pressure to do the right thing."

* * *

The N.R.C.'s own deliberations that weekend on whether to order a major evacuation of the region surrounding TMI-2 were largely focused on a narrow technical issue: Would the hydrogen bubble explode? When the bubble was initially discovered on Thursday night—officials at TMI-2 had noticed that the reactor coolant system was "soft" and not behaving as it would if the reactor were filled completely with water—there was no worry about the bubble's explosive potential. Hydrogen is explosive when mixed with oxygen, but pure hydrogen is not explosive at all. Since there was no free oxygen in the reactor available to mix with the hydrogen, the possibility of an explosion was dismissed. N.R.C. officials did not know on Thursday night, however, about the hydrogen explosion in the TMI-2 containment building that had occurred at 1:50 p.m. on Wednesday. This was not known at the Incident Response Center until Friday morning, when Mattson belatedly reported that on Wednesday afternoon there had been what he called a "funny blip" in the charts showing the pressure in the containment building— a "spike," indicating a sudden increase and then an immediate decrease in the pressure. Mattson told the commission that this was suggestive of a hydrogen explosion. Subsequent air samples from the containment building confirmed his surmise: they showed a deficit in the amount of oxygen normally present, some of the oxygen having been consumed when it combined with hydrogen to produce the explosion. Still, this was an explosion in the containment building, which is normally filled with air, and not in the reactor, which is normally filled with water and has no free oxygen available to support any explosion. Thus, news of a hydrogen explosion in the containment building merely confirmed the fact that hydrogen had been created in the reactor by chemical reactions early in the accident, and was not taken as an indication of any explosive potential inside the reactor. Hendrie himself, in a Friday afternoon telephone conversation with Governor Thornburgh, stated unequivocally that there was "no oxygen" in the bubble inside the reactor and hence no concern about an explosion.

Later on Friday night, however, Chairman Hendrie started to worry about the possibility that there might be a source of oxygen inside the reactor that could lead to a detonation of the hydrogen bubble. He was concerned, in particular, that a process known as radiolysis might be breaking down the water in the reactor into its component hydrogen and oxygen. Radiolysis can take place when radioactive materials are present in water because strong radioactive emissions can, in effect, dissolve the glue that bonds hydrogen and oxygen together to make water. Hendrie started making his own calculations of the rate at which radiolysis might be taking place and the rate, therefore, at which oxygen might be added to the bubble. The situation, as far as he could tell from his preliminary calculations, didn't look good.

Hendrie, who was at his home, started to make a series of late-night phone calls that he carried on into the early hours of Saturday morning. Shortly before 11 p.m. he called the Incident Response Center and told officials there about his calculations. He directed the N.R.C. staff to find experts who could make further calculations on the rate at which oxygen might be accumulating inside the bubble and assess the magnitude of any explosion that might occur if the hydrogen detonated. Hendrie also called the TMI-2 site and repeated his concern to N.R.C. officials there. At 2 a.m. on Saturday morning, he spoke to Mattson at the Incident Response Center and asked him to check into the oxygen addition rate and the potential for an explosion, and to confirm that his request for more calculations had been received and that the work was being done.

The technical evaluation of the bubble's explosive potential that was spurred by Hendrie's telephone calls continued throughout the weekend. It involved a frantic effort to pull together what Hendrie called his agency's "integrated intellects" aided by "most of the ranking world experts on the subject." The more N.R.C. delved into the potential for a hydrogen explosion, the worse the situation looked. Thus, by midmorning on Saturday, the commission began to worry that even if enough oxygen did not build up inside the bubble to produce an explosion, there might still be enough oxygen to make the bubble flammable. If the bubble caught fire, it was feared that this might cause a pressure surge inside the reactor that could further injure the already damaged core or damage the reactor vessel itself. Hendrie kept pressing for "an opinion from the hydrogen flammability crowd" on the amount of oxygen that it would take to produce a flammable mixture. Since Hendrie admitted "we haven't got our hands around the hydrogen—around the bubble problem from the standpoint of flammability," Commissioner Bradford kept asking whether there was "a sequence of events that could start anytime without warning" and leave little time to evacuate the surrounding population. Hendrie insisted that he "didn't think it's a very large possibility but you can't rule it out." Nor, he acknowledged, could one "rule out" the possibility that such a sequence of events might damage the containment chamber surrounding the reactor and allow a leakage path for large amounts of radioactive gas to escape from the plant into the neighboring countryside.

At noon on Saturday, Chairman Hendrie got a "general rundown" from Harold Denton on the conditions at the plant and told Denton about the concern at headquarters over the explosion and flammability problems with the hydrogen bubble. "My concerns in this area actually run two ways," Hendrie explained.

> One of them has to do with whether we may be already close enough to a situation where one ought to consider some further evacuation measures. And the other one is in moving the gas bubble around, if we get it out into the containment, I believe we're going to be flammable. That is, if it doesn't go in the [reactor] vessel and you discharge it into the containment, why I think that takes you well up into the flammable region.

Denton replied that these problems had not been

> high on my scale of concerns. You have heightened my worry about it. I guess I need—really, it is going to take a lot of assistance from back there, people looking at this, I guess, to get me and us up here up to speed on it.

At 3:27 p.m. on Saturday afternoon, all five commissioners and various assistants met with Roger Mattson at the Incident Response Center and reviewed the anxiously awaited estimates of the situation inside the bubble. Mattson had turned to Robert Tedesco and Saul Levine of the N.R.C. staff to derive this information, and they, in turn, had asked their own staffs to do various computations and had asked Robert Ritzman of Science Applications Incorporated, whom Levine had recommended, to do calculations as well. Ritzman had been the "expert on hydrogen" for the N.R.C. *Reactor Safety Study,* and Levine called him "the best man in the country." As for how much oxygen was in the bubble, Mattson told the commissioners, Ritzman's "number says two percent. It could be as high as three percent oxygen in the bubble." Ritzman also estimated that a flammable mixture would not be created until the oxygen level reached "eight to nine percent," Mattson said, and that an explosive mixture would not be reached until ten to twelve percent oxygen was present.

Other commission consultants, from such organizations as the Knolls Atomic Power Laboratory and the Idaho National Engineering Laboratory, concurred with the general assessment given by Ritzman, Mattson said. Since it was believed that the amount of oxygen in the bubble was increasing by less than one percent per day, N.R.C. officials felt they had a grace period of a few days before the threshold for flammability was reached, and even longer before the risk of an explosion materialized. The hydrogen problem, Chairman Hendrie concluded, had not been "put to bed" but "I think at the moment we have a reasonable basis for believing it is not a problem for Saturday night. Okay? It is several days out into the next week before—the best judgment is, before we hit the flammability limit, and the detonation limit is more than a week beyond that." Mattson said that "we just have to follow it, day by day. We're doing better on every calculation we do with each passing hour."

Transcripts of the session reveal some relatively relaxed discussions and general bantering about evacuation criteria. Then Hendrie told the group, "Yes, it's a useful thing to keep working along. I wouldn't—don't pull anybody off the hydrogen problem" (laughter) "but work along." Commissioner Kennedy stated, "I keep thinking, you know, we may not be as close to the edge of that precipice as it seems all the time to us." The chairman talked with Governor Thornburgh and told him, "We don't think that any precautionary evacuation at the moment is called for and, as I say, the situation, the plant situation is slightly better." He advised, however, that Pennsylvania authorities remain "on an alert status."

Sunday morning, once again, the numbers changed, and once again, the commission's attitude toward the necessity for a precautionary evacuation also shifted. (Four of the commissioners were at their offices in Washington, but Chairman Hendrie had gone to Three Mile Island to be on hand when President Carter arrived; the President was to visit the plant in an effort to calm public fears about the situation.) Dudley Thompson of the N.R.C. staff briefed the four commissioners on the latest developments early Sunday afternoon. There was, according to the most recent round of calculations, he told them, already about five percent oxygen in the bubble, not the two to three percent reported on Saturday, and the threshold level for flammability, instead of being eight to nine percent, was now thought to be five percent, or slightly less. A number of "quick recalculations" were being done, Thompson reported, though, "for all practical purposes, we've got to assume the mixture is flammable, but I don't think anybody is assuming right now that he thinks it's an explosive mixture." The numbers, Thompson stated, were provided to him by Darrell Eisenhut.

At 1:50 p.m. on Sunday, the commissioners received further word on the subject in a telephone call from Robert Budnitz, the deputy director of the Office of Nuclear Reactor Research. He told them that the N.R.C. had three groups of consultants doing calculations and that "we now understand what the flammability problems are with that stuff in the upper head [of the reactor vessel], and I'll give you the numbers, if I can find them." Budnitz reported that "we now think there's three to four percent oxygen, and the rest is hydrogen" inside the bubble, and the threshold at which the mixture would burn was when it reached four and eight-tenths percent oxygen. Such a mixture could burn quickly, in ten or twenty thousandths of a second, Budnitz told the commissioners, producing an enormous pressure pulse that might rupture the reactor vessel. "We might lose that vessel, which we can't afford," he said. "We're going to lose valves; we're going to lose seals; we're going to lose the pumps. We just can't stand that." These, he emphasized, were merely the effects of a rapid fire in the bubble. If the oxygen built up to the explosive level, which he now said could take place when the oxygen level reached ten percent (instead of the upper limit of twelve percent reported to the commission on Saturday), then "we're going to lose everything." Budnitz went on to discuss other complicated problems that might occur because of the buildup of oxygen in the bubble, and he observed, with some exasperation, "One thing I've found out, this agency needs chemists."

* * *

The potential explosion of the hydrogen gas bubble inside the reactor, and the potential for the bubble, even if it did not explode, to interfere with the cooling of the core, were the dominant but not the sole factors in the commission's deliberations on the need for a precautionary evacuation. In addition to these particularized concerns, the extensive, unprecedented damage that had occurred to the core of the Three Mile Island reactor created such an unfamiliar situation that, merely on the basis of general prudence in the face of large uncertainties, the commissioners wondered whether a precautionary evacuation might be justified. The N.R.C.'s generalized anxiety was underscored by what Roger Mattson told the commissioners on Saturday afternoon:

> Let me say, as frankly as I know how, bringing this plant down is risky. There's a not negligible risk in bringing this plant down. No plant has ever been in this condition, no plant has ever been tested in this condition, no plant has ever been analyzed in this condition in the history of this program.

Essential equipment then being used at Three Mile Island to cool the core, such as the main cooling pump that was circulating water through the core, was not designed to operate in what is known as the "postaccident environment." This equipment, located inside the reactor containment chamber, is designed to operate under normal conditions (low humidity, low temperature, and low radiation levels) during routine plant operation. As a result of the accident, however, this equipment had been exposed to a hostile new "environment" consisting of hot steam and murderously high radiation levels. Some of the equipment was even under several feet of water that had been dumped into the containment building from the reactor during the early stages of the accident. Some equipment—the so-called safety-grade devices—was specially designed to operate under these adverse conditions; special insulation, for example, had been put on some of the electrical devices to prevent steam or moisture from causing short circuits. The main cooling pump being used to cool the core, however, was "non-safety-grade" and its continued operation in the hostile postaccident containment environment could not be counted on. "Your pump is running in a condition it doesn't like to run in," Mattson told the commissioners. "You've got radiation in that containment with equipment that you're depending on now that's not radiation-qualified." He told the commissioners that there was a time period during which the equipment, though not designed to work under these difficult conditions, could be expected to stay in operation before its performance degraded until it failed completely. This grace period constituted an ill-defined "margin of safety" that Mattson said was "not large," though he suggested it was perhaps large enough for the commissioners to postpone their decision on a precautionary evacuation. He altered his previous recommendation for an evacuation, saying:

> I think, yes, my recommendation to evacuate I'd change today. When I made that recommendation . . . it was on the basis that you were releasing large amounts of radioactivity unfiltered up the stack with no apparent way to stop it and were allowing [the operators] to do it with no recourse. . . . I saw that release driving a quick decision on going to [the residual heat removal system] and the early indications were that the procedure they had in hand would fail and that the core would melt. I didn't have any choice but to make that recommendation. Almost an hour away from starting a core melt sequence, what else could I say?

Although he was now reversing himself and recommending that an evacuation decision be postponed, Mattson stipulated that "If the hydrogen explosion thing changes, I'll probably change my position here."

The commissioners were hardly put at ease by the quixotic evacuation recommendations they were receiving from their technical staff, and the ever-changing calculations relating to the hydrogen bubble added to their sense that the situation at the plant was not well understood—and possibly not well in hand. Commissioner Bradford protested to the other commissioners on Sunday that he would be "more comfortable" if a "lot of people" had been working for a "lot of years" on the type of technical problems then besetting the Three Mile Island plant, "instead of some very tired people who have been on it for twenty-four hours." Commissioner Gilinsky repeatedly raised the question of whether a precautionary evacuation might not be recommended simply as a kind of "insurance policy" that could be taken out while officials were in the process of gaining a better understanding of the prospects of bringing the plant safely under control. Gilinsky said that if he had "a friend in Harrisburg," which is about ten miles from the plant, "I guess I'd—I don't think I'd tell him to move, I'd tell him to keep close to his radio. . . . If you had somebody really close, you might tell him, if he didn't have to stick around, why, maybe he oughtn't to be there." Commissioner Ahearne, similarly, said on Sunday that he was predisposed to recommend an evacuation of those who, like Commissioner Gilinsky's hypothetical "friend," lived within a few miles of the plant.

Thus, according to the transcripts of the commission's closed-session deliberations on Saturday, March 31, and Sunday, April 1, the majority of the commissioners—Bradford, Gilinsky, and Ahearne—wanted to recommend a precautionary evacuation. Curiously, though, no such recommendation was ever made by the N.R.C. Late that Saturday night, and, of course, knowing nothing of what had been transpiring at the commissioners' closed meetings, I managed to get through by telephone to the commission. I spoke to Commissioner Bradford, who was then sitting in his office talking with Commissioner Gilinsky, and I asked why the commission was not urging further evacuation. Since, judging from the information I had received that day from the Incident Response Center, the commission could provide little assurance that the equipment necessary to keep the reactor under control would remain in good working order, how, I asked, could the commission fail to urge the precautionary evacuation of the population that would be most directly and most quickly affected by a large radiation release? Commissioner Bradford initially replied, tautologically, that this step was not being recommended because the commission hadn't decided to recommend it. When pressed further, he said "something more" was involved in the commission's deliberations that he was "not at liberty" to mention. Needless to say, this response provoked a stream of questions about the mysterious factor that was swaying the commission's decisions. Commissioner Bradford said no to all my questions, and reasserted that there was a "set of considerations" he could not discuss.

The sensitive issue that Commissioner Bradford would

not discuss on March 31, he explained several weeks later, was the fact that a majority of the commissioners were generally agreed on the need for precautionary evacuation, but they were reluctant to force the issue and override Chairman Hendrie. The matter was never formally put to a vote, Bradford acknowledges, and it would be hard to pinpoint the exact form of an evacuation recommendation that would have drawn three of the five votes on the commission. Nevertheless, Bradford says it was clear to him—and it is now well documented in the transcripts of the commission's closed meetings during those two days—that a majority of the commissioners did support an immediate general evacuation, at least for the region within a few miles of the plant. Chairman Hendrie steadfastly opposed this step, and the rest of the commissioners did not try to force a confrontation to break the impasse. Commissioner Bradford says he had the "inchoate feeling that the commission had to function as a unit" without the "luxury of dissent," and that it could not afford the "ill-feeling" that would be occasioned by a split on such a crucial issue. Such a recommendation required a stronger commission consensus than existed, Bradford said, although he admits, "I'm not sure whether these are good reasons or creditable reasons in retrospect, but they were my reasons at the time."

* * *

Events outran the N.R.C.'s decision-making paralysis, and an order for a precautionary evacuation became a moot issue for two reasons. The first was that the neighboring population was not much inclined to wait around for further guidance from Washington in the light of what they had already been told. By the tens of thousands, the residents of the neighboring communities packed their belongings, assembled their children, locked their homes, and fled. The inhabitants of the surrounding area—whom Commissioner Ahearne described in one closed commission meeting as three quarters of a million "tense" people—had been exposed to three days of sanguine official reassurances, interspersed with shocking and unexpected contradictions. They were told on Wednesday, by Metropolitan Edison Vice President Jack

Herbein, that there had been some "minor fuel failure," and were reassured by the N.R.C. on Thursday that the situation was well in hand with a satisfactory recovery operation under way. The next morning, Friday, March 30, they were told of "an uncontrolled radiation release" from the plant and were advised to stay inside their homes. By Friday noontime, pregnant women and children had been advised to evacuate, with the rest of the population expected to sit by awaiting further instruction. On Saturday, instead of instruction, they received a series of alarming reports telling them of a hydrogen bubble and its explosive potential. (Anyone familiar with the fate of the German airship the *Hindenburg*, which was filled with hydrogen, would have known what a hydrogen explosion was like.) They were also told by Chairman Hendrie that "consideration" might have to be given to a more general evacuation, one that would require all the people who lived as far away as twenty miles from the plant to leave their homes. Many people did not bother to wait for official recommendations, especially those in the communities closest to the plant, but departed during the weekend. Some one hundred and forty-four thousand people are estimated to have left the area, including sixty percent of the people who lived within five miles of the plant. Some sixty-five percent of the pregnant women and small children living within fifteen miles of the plant left. Evacuees, on the average, traveled a distance of one hundred miles seeking safety from the accident, and stayed away from their homes for several days. Postaccident studies have shown that this massive disruption of the everyday lives of the people in the surrounding region had a profound psychological effect on many of them, with pregnant women suffering the most acute stress.

By Sunday night, there was another development that helped to moot the evacuation issue. Senior N.R.C. officials, including Chairman Hendrie and Roger Mattson (who had both gone to the site earlier in the day), marched shamefaced that evening into a meeting of industry experts who had assembled to advise on the recovery operations. The N.R.C. officials admitted that they had made a major technical error; they had belatedly learned, they said, that a hydrogen explosion inside the reactor was impossible because there was, in fact, *no oxygen* in the bubble at all. Earlier that day, Mattson, who had been relying on estimates of an oxygen buildup provided by a variety of N.R.C.'s experts and consultants, had gone to the site, where he had talked with Victor Stello. Mattson recalls that "the first words out of Stello's mouth" were that "there is something wrong" with the oxygen calculations. By around three on Sunday afternoon, Stello and Mattson had checked back with N.R.C. experts, only to learn that they had changed their minds: the hydrogen-explosion problem had vanished. More refined calculations had shown that the process of radiolysis inside the reactor was not actually producing a net output of oxygen. As the radiolysis proceeded, the hydrogen and oxygen in some of the water molecules were indeed being separated. However, because of all the hydrogen in the system, the oxygen was recombining with hydrogen to form water. In fact, hydrogen is normally added to many reactors so that any oxygen present will combine with it to form water rather than accumulate in the system. The hydrogen bubble had, as it were, an inherent self-defense mechanism against an oxygen buildup—it would convert any oxygen that evolved into water—and the early calculations by the experts had not adequately accounted for this phenomenon. The industry experts who met with Mattson and N.R.C. officials on Sunday night berated the government's analysts as "dummies," Mattson said, and some members of the industry team, such as Ed Zebrowski, the co-chairman of the group, accused the N.R.C. of not knowing "sophomore-level nuclear engineering." N.R.C. officials replied that this was an overstatement, and that, while their calculations had been mistaken, this was the result of an effort to be conservative; to err on the safe side.

* * *

The bubble was not only defused and its explosive potential contradicted by revised calculations of its oxygen content, but it also, unexpectedly, disappeared altogether. While the N.R.C. had been focusing on the problem of a large, growing bubble in the top of the reactor, the hydrogen gas present inside the system was being steadily removed in two straightforward ways by the Three Mile Island operators. Some of the hydrogen was being bled out of the reactor through the normal purification system, which takes a small stream of water out of the reactor and removes dissolved gases and foreign matter. In addition, a vent valve in the reactor's pressurizer tank was being used to transfer hydrogen directly into the containment chamber. The N.R.C., however, instead of acknowledging and relying upon these two mundane techniques, had been frantically reviewing a variety of convoluted schemes for getting rid of the bubble. All of the schemes entertained by the N.R.C. as part of what Chairman Hendrie referred to as the "head scratching" on "how the hell we get rid of" the hydrogen bubble overlooked the two simple approaches, known to industry experts, that easily and quickly removed the bubble. By Sunday night, Roger Mattson conceded, the hydrogen bubble "disappeared rather suddenly." Curiously, the removal of the hydrogen was accomplished with non-safety-grade equipment—the purification system and the small vent valve—that fortuitously survived the adverse conditions inside the containment chamber. In dealing with the hydrogen bubble, design foresight, which would have led to the incorporation of high-reliability safety equipment for this purpose, played a lesser role than simple good luck.

The N.R.C.'s concern about the hydrogen bubble and its explosive potential had created a national and international news story, and reports about the status of the bubble were broadcast by the news media with all the attention given to the scores from the World Series. On Saturday, Jack Herbein of Metropolitan Edison had announced that the size of the bubble had been sharply reduced and that the crisis was over, but Harold Denton, at another news conference an hour later, had declared that the crisis was not over, his contradiction adding to local concern and confusion about the situation at the plant. Even after Sunday afternoon, when the N.R.C. had realized its blunder in postulating the presence of a growing amount of oxygen in the bubble, it issued no announcement to disclose this mistake to the public. It was not until two days later—April 3, 1979—that the N.R.C. announced that "the bubble has been eliminated for all practical purposes." This announcement, which was made by Denton, did not include any candid account of the basic mistake that the N.R.C. had made in its prior statements about the risk of an explosion. Instead, when Denton was asked why the bubble had gone away, he replied, "I think it was a little bit because of our actions and maybe a little bit of serendipity." He thus made it appear that the N.R.C. deserved credit for removing the bubble, although this had been achieved two days earlier by Metropolitan Edison, using simple methods suggested by Babcock & Wilcox.

* * *

During the five-day period after the N.R.C. was notified of a general emergency at TMI-2, senior N.R.C. officials issued various public statements diagnosing the plant's conditions and made a number of internal judgments, not made public at the time, that formed the basis for the commission's handling of the event. "There never was a very close correlation between what was happening to the reactor and what N.R.C. thought was happening," Commissioner Bradford observed after the accident. Referring back to his hypothetical scale of one to ten for rating the severity of reactor accidents, on which five was the threshold for ordering an evacuation, he noted that the perceived danger during the Friday-to-Sunday period was well above the threshold, whereas the actual danger was well below it, precisely the inverse of the relationship between appearance and reality that had prevailed on Wednesday and Thursday. It would be comforting to think, he observed, that the N.R.C. staff would always happen to err on the side of caution, as they did in the weekend analyses of the hydrogen-bubble problem. Given the ambiance of pronounced technical confusion in which the staff's judgments appear to be made, however, he concedes that there is little basis for thinking that the N.R.C. staff's mistakes would follow such a fortuitous pattern. They certainly did not "err on the side of caution" on Wednesday the 28th, he noted. The combination of staff technical misjudgments and the commission's own decision-making paralysis, Bradford conceded, constituted an embarrassing public display of the level of competence that the N.R.C. brings to bear on the regulation of the nuclear-power industry. "The N.R.C.'s crisis management," Bradford observed, "was clearly a fiasco."

# Epilogue

Scientists prefer to see important tests and experiments—especially potentially hazardous ones—performed under carefully controlled conditions. Valuable data are occasionally acquired, however, in unplanned and less than scientific ways. The Three Mile Island accident, which provided an unscheduled test of the safety precautions in a typical large nuclear-power station, is a case in point. TMI-2 was owned by a middle-sized electric utility company, operated by a crew of above-average competence, equipped with a standard pressurized-water reactor with standard safety features and a standard large control room, and licensed under the customary safety rules of the Nuclear Regulatory Commission. It provided, in other words, a representative "test case." The maintenance crew that reported for the March 27, 1979, overnight shift had no cause to think of itself as a professional safety-review committee, yet its inadvertent disruption of the plant probed TMI-2's safety precautions as well as any stern examiner might have. By the time the accident was over five days later, it had provided an unprecedented opportunity for assessing the quality of the equipment, procedures, and people responsible for protecting the American public from nuclear radiation accidents.

The accident—and the corresponding test of the safety precautions governing the plant—proceeded in distinct phases. The opening phase, involving the maintenance workers as they went about their routine cleanup chores on the plant's water-polishing equipment, was a test of the management of the plant. The management was responsible for organizing the maintenance work and for seeing to it that equipment in the plant was designed so that it could be maintained without jeopardizing the safety of the reactor. Plant management failed to fulfill this responsibility. Metropolitan Edison officials had ignored dramatic earlier warnings that routine maintenance on the polishers could totally disable the main feedwater system at the plant, and had failed to change procedures or equipment designs to forestall a recurrence of this difficulty.

Once the feedwater system had been disrupted, the next phase of the accident tested the automated safety equipment in the plant and whether it would be capable of stabilizing the reactor and keeping it cooled. It was not. Designers had skimped on the installation of sophisticated automated safety devices and controls that were needed for this purpose, and part of the nominal plant safety equipment—the emergency feedwater system—had been disabled a few days earlier during testing and not restored to service (another reflection on overall plant management).

As the accident continued, more equipment was called upon to work, such as the pressure-relief valve, but it malfunctioned and considerably worsened plant conditions, turning an otherwise routine shutdown into a serious loss-of-coolant accident. Since automated equipment to correct a stuck-open relief valve was not available, the developing accident became a test of whether the operators could handle this situation. They could not. They were not given the instruments required to diagnose the problem, and the superficial training they had received left them unprepared to cope with a novel and complex emergency situation. Even the most senior members of the plant's management failed to organize appropriate remedial action.

When a General Emergency was finally declared, and the N.R.C. was called in to help stem the crisis, the competence of this key organization was given a sudden and very searching examination. N.R.C. officials, all the way from the junior inspectors who were sent to the plant to the chairman of the agency, performed with marked ineptitude.

The TMI-2 crisis was a complicated and protracted event, and by the end of the fifth day after it began, it had tested a very large number of individuals, plant components, safety systems, instruments, communications links, emergency procedures, and much more. Having touched so many separate aspects of the plant and having involved so many people, the accident constituted a comprehensive general examination of just about everything that went into assuring the safety of TMI-2. The resulting scorecard, tallied with any objectivity, would not show a passing grade in any category. Even the ultimate success in heading off a core meltdown hardly qualifies as a planned and carefully executed maneuver. Forestalling a meltdown was dependent on closing a key valve, which was done at 6:18 a.m. at the suggestion of Brian Mehler, who was not part of the crisis-management team and who acted without the knowledge of, or on the basis of a directive from, plant management. He, like everyone else at the plant, was unaware that the core at that point was on the verge of melting down. Keeping the core cooled for the rest of the day depended on the happenstance that the throttled emergency cooling pumps, which were turned on and off throughout the day in nearly random fashion, luckily allowed enough water to be delivered to the reactor. These actions were undertaken by the plant staff without any understanding of the state of the reactor or of the necessity for taking these steps. Plant officials were actually focusing their attention on other priorities, and had the central maneuver attempted by plant officials on the 28th actually succeeded—that is, had they managed to put the reactor on the normal shutdown cooling system—an entirely different outcome might have occurred. Completion of the strategy they adopted would have meant pumping intensely radioactive water from the reactor outside the containment building, thereby threatening a massive radiation leak outside the plant. The prevention of a meltdown—which the TMI-2 staff did not even recognize as an actual threat—depended largely on inadvertent actions.

* * *

In the weeks immediately following the accident, various official investigations into its causes and implications were announced. President Carter set up a special commission, chaired by the mathematician John Kemeny, the president of Dartmouth College. House and Senate committees and subcommittees undertook their own investigations. The N.R.C. announced that it would form a special study group and that an existing branch of the agency (the Office of Inspection and Enforcement) as well as an ad hoc "Lessons Learned" task force would also study the accident and its ramifications. Babcock & Wilcox, Metropolitan Edison, and others with a direct involvement—and potential legal liability—undertook their own reviews, and industry organizations, such as the Electric Power Research Institute, initiated parallel, independent investigations. The published results of these studies provide voluminous details about the accident, about the performance of all the principals, and about the general policies and practices of the N.R.C., the agency that had approved everything about the plant and allowed it to go into commercial operation. There is wide agreement on most of the factual details about what happened during the accident, but there are conflicting assessments of who was responsible for its occurrence and its ultimate severity. Babcock & Wilcox, the designers of the reactor, blame the operators, while Metropolitan Edison and its operators blame the designers, whom they are suing for five hundred million dollars. The N.R.C.'s Office of Inspection and Enforcement blames the operators and the operators' "mind-set"—that is, the unfortunate interpretation they made of their control-room instrument readings, which led them to shut off the emergency cooling pumps that should have been left on. The President's commission, on the other hand, points the finger at the N.R.C.'s own "mind-set": the general complacency in the agency that permitted a plant with so many pronounced safety defects to go into operation. "With its present organization, staff, and attitudes," the presidential panel concluded, "the N.R.C. is unable to fulfill its responsibility for providing an acceptable level of safety for nuclear power plants." The

N.R.C.'s own Special Inquiry Group, composed chiefly of its own staff but directed by the Washington attorney Mitchell Rogovin, reached a similar conclusion, and its final report is a multivolume set of complaints about how the agency fails to carry out its basic mission to assure nuclear-plant safety.

The harsh criticisms directed at the N.R.C., together with the agency's self-criticism, have yet to bring about more than a few limited reforms. The presidential panel recommended that the five-member commission be abolished and that a new agency be established, headed by a strong administrator uncommitted to past N.R.C. policies. In December 1979, President Carter decided instead simply to ask the members of the existing commission to change places. He removed Joseph Hendrie, whose performance during the crisis drew heavy public criticism, as chairman, but allowed him to remain as a commissioner. At the same time, President Carter promoted Commissioner John Ahearne to the chairmanship. (A few months later, Ahearne delegated to Hendrie the responsibility for heading the agency's response to future accidents on the grounds that Hendrie was the most qualified member of the commission for this assignment.) President Reagan, shortly after taking office, undid Mr. Carter's limited changes by reappointing Hendrie chairman of N.R.C. for the remainder of his term, which ended in June 1981. The current chairman of the agency, Nanzio J. Palladino, has yet to announce any plans for upgrading N.R.C.'s regulatory programs.

Other bureaucratic reshufflings have taken place, presumably to improve the federal government's performance in the event of a future General Emergency. Some of the crisis-management responsibilities that had been assumed by the N.R.C. were formally shifted to the federal Emergency Management Agency. (The N.R.C., however, "loaned" seven of its own staff members to this agency, so that there will be no major change in the personnel handling future crises.) No general overhaul of its staff has been effected by the N.R.C. in the wake of the TMI-2 accident. The N.R.C.'s Special Inquiry Group—which was the most suitable review group to judge the performance of N.R.C.'s personnel in the licensing of the plant and in the handling of the emergency—made no findings about the competence or fitness of individual N.R.C. staff members for the responsibilities they hold. None of the individuals responsible for the many serious mistakes that led to the crisis have been replaced, and the senior management team of the N.R.C., which is still headed by Harold Denton, is virtually the same now as before the accident.

The N.R.C.'s staff assignments have shifted somewhat since the accident, of course, because one of the agency's current concerns is the massive cleanup operation at TMI-2. The crippled plant contains immense quantities of radioactive material. Some of it remains concentrated in the "rubble-ized" reactor core, but there is also the irksome and potentially hazardous problem of a million gallons of radioactive water that floods the bottom of the containment building and fills various makeshift storage tanks. Despite all the time that has passed since the accident, the N.R.C. and the nearly bankrupt Metropolitan Edison Company have yet to organize a satisfactory plan for disposing of this ultrahazardous material (and no contingency plan has been prepared for continuing with the cleanup, should Metropolitan Edison go bankrupt). Prior to the accident, the federal government had no concrete plan for disposing of the radioactive wastes from the commercial-nuclear-power program, and officials are now hard pressed to figure out what to do with the deadly debris created by the accident, which still contaminates the TMI-2 facility.

Efforts to clean up the plant have not gone smoothly, and various leaks and spills, while not great in quantity, have been enough to cause continuing anxiety to the neighboring population, which was already badly traumatized by the accident itself. Technical "glitches"—to use Harold Denton's term—have repeatedly marred the effort and the N.R.C. has not demonstrated, by its own performance in supervising the cleanup, any large measure of newfound competence. The N.R.C. approved a plan to vent the radioactive krypton gas from the containment building, for example, and then, when the venting was started, it had to be suspended after only four minutes because of a high-radiation alarm. Denton, who has admitted that the agency had a "complacent" attitude before the accident, described the snag in the venting effort in these terms: "I told my family coming up here [to witness the venting] that based on past experience, there was only a fifty-fifty chance that this star-crossed plan would work." Denton added that "it obviously wasn't an adequate procedure to start with.... I am concerned, along with the people here, that there is always a last-minute snag." What will become of all the radioactive material at TMI-2, and whether it can be disposed of without harm to the local population, are unresolved questions. At the moment, the plant just sits there, no longer a nuclear-electric-generating facility—and herald of the "bright and shining option" of a nuclear-powered future envisioned at its dedication ceremonies in September, 1978—but a de facto radioactive-waste-storage dump.

Under Harold Denton's general supervision, the agency staff has prepared an "action plan" that describes in detail what the N.R.C. is attempting to do in response to the accident to improve the safety of nuclear plants. Some small steps have already been taken. The N.R.C. now requires, for example, that an engineer be assigned to each shift to help the operators cope with serious accidents. Some minor modifications have been made to plant control rooms, such as additional instruments, but one of the consultants the N.R.C. retained to review the weaknesses in the "human engineering" of current control rooms said in a recent interview that the N.R.C.'s "improvements" have actually made things worse: they have added more clutter to already overcrowded and still badly designed control rooms. A dramatic improvement in nuclear-plant designs and safety practices has yet to be attempted, since the N.R.C. "action plan" is mostly a catalogue of the agency's intentions to "study" numerous issues and do "appropriate research." As some of the items on the list have been included in similar lists for the last fifteen years, the schedule on which needs for additional safety improvements will be determined, and then brought about, is uncertain. The backlog of major unresolved safety problems is so large that it will obviously take several years to implement necessary safety improvements—perhaps, even with a high federal priority, the better part of a decade.

What is to be done in the meantime with the seventy-one nuclear plants now operating in the United States under the

N.R.C.'s auspices? The overall safety of these plants—a worrisome question in the aftermath of the Three Mile Island accident—was curiously sidestepped in the official reports on the accident. The presidential commission, for example, in a promisingly forthright paragraph on page 24 of its report, stated that it "had to face the issue of what should be done in the interim with plants that are currently operating and those that are going through the licensing process." Immediately after raising this urgent question, however, the presidential commission quickly turned away from it. Nowhere in the one-hundred-and-seventy-nine-page report did it again, except in passing, discuss the problems posed by the continued operation of these facilities, some of which are uncomfortably close to the New York, Boston, Chicago, Sacramento, and other metropolitan areas. Its silence on this sensitive subject was determined, it would appear, before the outset of its work. President Carter, by means of Executive Order 12130 signed on April 11, 1979, sharply restricted the investigation of the group to the Three Mile Island accident. The panel was explicitly told to evaluate the N.R.C.'s performance "as applied to this facility." The Carter administration, strongly committed to nuclear power as a part of the President's "comprehensive national energy policy," evidently had no desire to ask pertinent larger questions about the overall safety of the existing plants, about whether there really was a basic necessity for a major commitment to nuclear power, and other questions that could conceivably bring the commission's conclusions into conflict with established federal policies. President Carter, who had some technical background in nuclear power as well as some political sophistication, knew what questions *not* to ask. The Reagan administration, even more strongly committed to nuclear expansion, has similarly avoided these questions.

The troublesome questions about the safety of the seventy-one operating U.S. nuclear plants will not go away simply because the administration and the N.R.C. do not want to discuss them candidly. Some decision needs to be made on whether the currently operating plants are "safe enough," whether urgent safety repairs on some of them might be needed, and whether some of them—especially those with poor safety records and those located in very heavily populated areas—might not best be shut down.

In 1975, the N.R.C. published an eleven-volume *Reactor Safety Study,* the so-called Rasmussen Report, which presented estimates of the extremely remote possibility of calamitous reactor accidents. This mammoth study was officially regarded as a stunning accomplishment that eliminated all reasonable doubts about nuclear-plant safety. Scientists were not so sure about its results, however, and their criticisms prompted the N.R.C. to organize its own review of the study. The N.R.C. review group concluded in September 1978 that the study's accident-probability estimates lacked a sound technical foundation. (The review group noted that the "arbitrariness" of some of the Rasmussen Report's methods "boggles the mind.") In January 1979, the N.R.C. withdrew its official endorsement of the Rasmussen findings and repudiated the study's numerical estimates of accident probability as not "reliable." N.R.C. Chairman Hendrie explained to Congress at the time that this did not diminish his confidence in nuclear-plant safety, because the basis for the assurance of safety was the commission's regulatory framework, he said, rather than some abstract study. According to the N.R.C.'s own investigation of the TMI-2 accident, however, and according to the findings of the President's commission, the existing regulatory framework provides little basis for confidence in the safety of current nuclear plants. With the Rasmussen Report repudiated, and the regulatory framework discredited by the accident, there remains only the most uncertain basis for confidence in the safety of currently operating nuclear plants.

The overall statistics that summarize the operating history of the commercial nuclear-power program reinforce the sense of unease. The U.S. nuclear power industry has accumulated slightly more than six hundred reactor-years of operating experience. Included in this experience are three close calls: the 1966 partial meltdown of the Fermi breeder reactor, the 1975 Browns Ferry fire, and the 1979 accident at TMI-2. Statistically, then, the established frequency for accidents of this degree of seriousness is about one in every two hundred reactor-years of operation. With currently licensed reactors, the United States now achieves two hundred reactor-years about every three years, and it will accumulate reactor-years more quickly if the dozens of nuclear plants now under construction are allowed to go into commercial operation. Thus, on the basis of the available record, with the current level of safety, a major reactor accident—one comparable in severity to those at Fermi, Browns Ferry, and TMI-2—should be expected about every three years. What the odds are for even more serious accidents that do not stop short of a meltdown, no one can say with assurance. According to the conventional wisdom in the nuclear-power industry, as expressed only a few years ago, meltdown accidents were so improbable as to be deemed "incredible" events. The industry said that there was less than a "one in a million chance" of a meltdown per reactor-year of operation. The experience of TMI-2, which came within thirty minutes to an hour of a meltdown only three months after it began commercial operation, does not lend much support to such an optimistic assessment. Some of the safety modifications the N.R.C. may ultimately require of the industry could improve the odds against a meltdown, although neither the schedule for these improvements nor the magnitude of the increased safety margins can be defined very precisely. On the other hand, as existing plants age and as more units come on line, new problems may offset the hoped-for improvements, although to an extent that remains correspondingly unclear. The TMI-2 accident was a surprise, and what other surprises remain in store no one can say, least of all the N.R.C. The industry is confident that the problems have been corrected; other observers are less sanguine.

A few months ago, some of the members of the N.R.C.'s Special Inquiry Group, which studied the Three Mile Island accident on behalf of the commission, got together for an informal luncheon in Washington. The discussion was "very gloomy," according to Mitchell Rogovin. In his judgment, the N.R.C. has done "virtually nothing" to carry out the recommendations that he and his colleagues made in January 1980 for improving the regulation of the American nuclear-power industry. Rogovin said he was "awfully sad" that instead of adopting a better regulatory program, the N.R.C. remained preoccupied with the creation of one that would simply license nuclear plants faster—"expediting the things that made us uncomfortable." Changes in the licensing process proposed by the N.R.C. in March 1981—which would cut back on the opportunities for the public to question plant safety—were "very troubling," he said. The American public has "an extraordinarily short attention span," and so pressure on the N.R.C. to improve plant safety has greatly diminished in the two years following the accident at Three Mile Island, he observed, adding, "People worried about the Americans held captive in Iran, but they have overlooked the fact that they, too, are hostages, to the nuclear plants operating around the country. No one would question whether there will be another accident. It's merely a matter of when."

# Index